Algebra and

on the

HP 48G/GX

by Dan Coffin

Grapevine Publications, Inc.

P.O. Box 2449

Corvallis, Oregon 97339-2449 U.S.A.

Acknowledgments

The terms "HP 48" and "48" are used for convenience in this book to refer to the HP 48GX and the HP 48G, the registered trade names for the handheld computing products of Hewlett-Packard Co. For their excellent programming tips and for allowing the reprinting of their programs, the author gratefully acknowledges Jim Donnelly (FMPLT, INPLOT and UDFUI) and Bill Wickes (→RPN, SDET, SMMULT STRN, and slightly modified versions of SA→, →SA, SCOF, SMADD, SMSMULT, SMSUB). Hugs and kisses in abundance to Bonnie and Ian for their unflagging love, support and encouragement throughout this project.

Printed in the United States of America
ISBN 0-931011-43-4

First Printing – September, 1994

CONTENTS

0. START HERE

What Is This Book?

This book is to help you use the HP 48G or HP 48GX calculator to improve your understanding and increase your enjoyment of the collection of mathematical topics usually grouped under the names of Algebra and *Pre-Calculus*. You may be a student currently enrolled in a Pre-Calculus course, a student in a Calculus course (which builds upon Pre-Calculus topics), or a student of lifelong learning who has an opportunity to learn (or re-learn) something new and useful.

This book organizes the material much as a standard text does. Chapters are divided into topics. Topics are divided into examples, each of which demonstrates how to use the HP 48 to solve and illuminate a problem of the kind you are typically asked to solve in a Pre-Calculus course. Occasionally, the examples make use of programs that extend the capabilities of the HP 48. The full listings for these programs are included (in alphabetical order) in the Appendix.

If you are currently a student in a Pre-Calculus course, please note that this book isn't meant to *replace* your textbook. In general, this book makes no attempt to rigorously justify the techniques and concepts used in problem-solving as does a standard text. Also, there may be topics treated in greater depth in your textbook than in this book (or vice versa).

What Do You Need to Know Before Using This Book?

You should have a basic working knowledge of your HP 48, including:

- Performing basic arithmetic calculations
- Navigating menus
- Entering alphabetic characters
- Storing, recalling, and using variables
- Entering and using lists, algebraic expressions, and programs

(If you need a quick review, stop here now and work through the *Quick Start Guide* or the first 3 chapters of *An Easy Course in Programming the HP 48G/GX*.) The only other things you need are a basic understanding of algebra, access to an HP 48G/GX calculator, and a willingness to explore Pre-Calculus mathematics.

1. EXPLORING FUNCTIONS

Linear Functions

Mathematically, a *function* is a process that accepts certain *inputs* and generates corresponding *outputs*—exactly one output for each input. A simple kind of function is the *linear* function, so-named because its graph is a line. Its slope-intercept form is $f(x) = mx + b$; m is the line's slope, and b is the y-intercept.

Example: Plot the linear function, $f(x) = mx + b$.

1. The HP 48 can plot a function when the *only* undefined variable is the independent variable (that's x in this case). So you must give m and b numerical values before attempting to plot the function. Use $m = 2$ and $b = 1.4$: From the stack, press $\boxed{2}$ $\boxed{'}$ $\boxed{\alpha}$ $\boxed{\leftarrow}$ $\boxed{M}$ $\boxed{\text{STO}}$ and $\boxed{1}$ $\boxed{\cdot}$ $\boxed{4}$ $\boxed{'}$ $\boxed{\alpha}$ $\boxed{\leftarrow}$ $\boxed{B}$ $\boxed{\text{STO}}$ to store the values for the coefficients.

2. Press $\boxed{\rightarrow}$ $\boxed{\text{PLOT}}$ to begin the **PLOT** application. Then press $\boxed{\text{DEL}}$ $\boxed{\blacktriangledown}$ $\boxed{\text{ENTER}}$ to reset the various plot parameters to their default values.

3. To change the content of a field in an input screen, you move the highlight to that field (using $\boxed{\blacktriangle}$, $\boxed{\blacktriangledown}$, $\boxed{\blacktriangleright}$, and $\boxed{\blacktriangleleft}$), then enter the information. Some items you can type; others you select via **CHOOS** or **✓CHK**. Now change the plot type to **Function**, if necessary.

4. Move the highlight to the **EQ:** field; enter the expression **'m*x+b'**: $\boxed{'}$ $\boxed{\alpha}$ $\boxed{\leftarrow}$ $\boxed{M}$ $\boxed{\times}$ $\boxed{\alpha}$ $\boxed{\leftarrow}$ $\boxed{X}$ $\boxed{+}$ $\boxed{\alpha}$ $\boxed{\leftarrow}$ $\boxed{B}$ $\boxed{\text{ENTER}}$. Note that only the right-side of the function need be entered. Set the **INDEP:** field to (lower-case) **X** (note that no **' '** are needed here): $\boxed{\alpha}$ $\boxed{\leftarrow}$ $\boxed{X}$ $\boxed{\text{ENTER}}$.

5. As for the part of the plot to be displayed, the defaults for **H-VIEW** and **V-VIEW** are adequate.

6. Press **ERASE** **DRAW**.

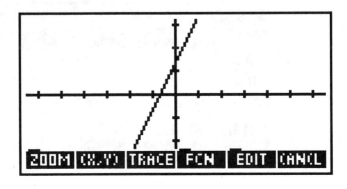

Example: Repeat the previous example with $m = 2$, but vary b with the values $-2, -.5, 1$, and 5. Plot each graph without erasing the previous ones.

1. Use CANCEL to return to the **PLOT** input screen, then NXT **CALC** to use the stack from within the application.

2. Press VAR to go to your VAR menu, then type the first desired value of b, 2 +/-, and press ← **B** (or, equivalently, ' **B** STO).

3. Press ← CONT **OK** (to get back to the **PLOT** input screen), then NXT **DRAW** (don't use **ERASE**; you want to see the plots together).

4. Repeat steps 1 through 3 for the other values of b ($-.5, 1$, and 5). The figure below shows all five lines plotted on the same graph.

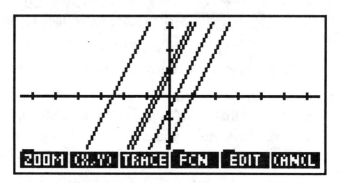

Example: Repeat the previous example, with b fixed at -1, but then vary m (use $-2, -0.5, 1$, and 5). And try using the FaMily PLoT (**FMPLT**) program (if you have previously entered it—see page 281). Here's how:

1. Exit the **PLOT** application with CANCEL. Then, from the stack display, type α α F M P L T ENTER.

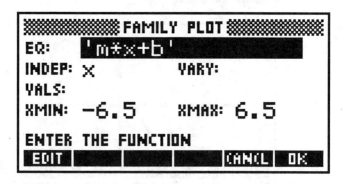

2. To store -1 in *b*, the procedure is the same as in the built-in **PLOT** application: NXT **CALC** VAR 1 +/- ⟵ **B** , then ⟵ CONT **OK**

3. The function displayed in the **EQ:** field is correct. Enter the independent variable, **X**, in the **INDEP:** field and the variable whose effects you wish to study over the several simultaneous plots (**M**) into **VARY:** field: ▶ α ⟵ X ENTER α ⟵ M ENTER.

4. Enter a *list* of its desired values into **VALS:**: ⟵ {} 2 +/- SPC • 5 +/- SPC 1 SPC 5 ENTER.

5. Leave the **X-MIN:** and **X-MAX:** as they are and press **OK** . The function is plotted four times, once for each value of *m* (which you'll see displayed as each line is drawn):

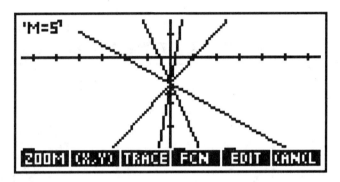

Example: Before you plot, it helps to know the range of a function. To easily find the range of, say, $f(t) = \dfrac{2}{3}t - \dfrac{5}{3}$, with $t \in \{-3,-2,-1,0,1,2,3\}$:

1. (Use CANCEL to exit **FMPLT**) Enter the right side of the function onto the stack: ⟵ EQUATION 2 ÷ 3 ▶ α ⟵ T – 5 ÷ 3 ENTER.

2. Enter the domain list: ⟵ {} 3 +/- SPC 2 +/- SPC ...etc., ENTER, And store the domain in '**t**': ' α ⟵ T STO.

3. Now just *evaluate* the function: EVAL. And rationalize the decimals in the result: 1 0 α α F I X ENTER ⟵ SYMBOLIC NXT **→Q** .

Result: A list of range values corresponding to the domain values:

{ '-(11/3)' '-3' '-(7/3)' '-(5/3)' '-1'
  '-(1/3)' '1/3' }

Quadratic Functions

Quadratic functions are functions of the form $f(x) = ax^2 + bx + c$.

Example: Practice more now with the **FMPLT** program, by exploring the effect of varying each coefficient (a, b, c) of a general quadratic. Remember to store values in the two coefficients not being varied:

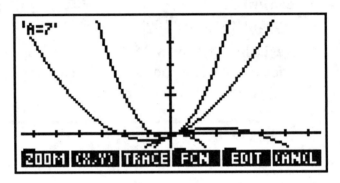

Varying a ($a = \{$ −5 −1 2 7 $\}, b = 4, c = -1$)

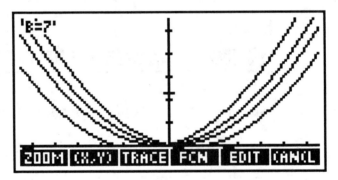

Varying b ($a = 3, b = \{$ −5 −1 2 7 $\}, c = 1$)

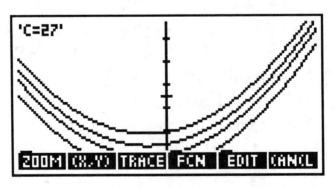

Varying c ($a = 3, b = 5, c = \{$ −20 −1 12 27 $\}$)

The *discriminant* of a quadratic is given by $d = b^2 - 4ac$. What does the sign ($\pm$ or zero) of the discriminant, d, indicate about the function's graph?

Example: Graph cases where $d = -4$, $d = 0$, and $d = 12$.

1. Arbitrarily let $a = c = 1$: $\boxed{1}\boxed{'}\boxed{\alpha}\boxed{\leftarrow}\boxed{A}\boxed{STO}\boxed{1}\boxed{'}\boxed{\alpha}\boxed{\leftarrow}\boxed{C}\boxed{STO}$.

2. By rearranging the discriminant equation, $b = \sqrt{4ac + d} = \sqrt{4 + d}$. Enter that equation for b: $\boxed{\alpha}\boxed{\alpha}\boxed{S}\boxed{T}\boxed{D}\boxed{ENTER}\boxed{\leftarrow}\boxed{EQUATION}\boxed{\sqrt{x}}\boxed{\leftarrow}$ $\boxed{()}\boxed{4}\boxed{MTH}$ **LIST** **ADD** $\boxed{\alpha}\boxed{\leftarrow}\boxed{D}\boxed{ENTER}$.*

3. Store the list of values for d into $'d'$: $\boxed{\leftarrow}\boxed{\{\}}\boxed{4}\boxed{+/-}\boxed{SPC}\boxed{0}\boxed{SPC}$ $\boxed{1}\boxed{2}\boxed{ENTER}\boxed{'}\boxed{\alpha}\boxed{\leftarrow}\boxed{D}\boxed{STO}$.

4. Press $\boxed{EVAL}$ to compute the resulting list of values for b.

5. Press $\boxed{'}\boxed{\alpha}\boxed{\leftarrow}\boxed{B}\boxed{STO}$ to store that list into $'b'$.

6. Use FMPLT to plot functions with $a = 1$, $b = \{0, 2, 4\}$, and $c = 1$: $\boxed{VAR}$ (then $\boxed{NXT}$ or $\boxed{\leftarrow}\boxed{PREV}$ as needed) **FMPL** to start the program. Then enter the quadratic function ($'$a*x^2+b*x+c$'$) into the **EQ:** field, if it isn't already displayed from the previous example. Make sure that the **INDEP:** variable is X and the **VARY:** variable is b. With the **VALS:** field highlight, press $\boxed{NXT}$ **CALC** $\boxed{\alpha}\boxed{\leftarrow}\boxed{B}\boxed{ENTER}$ **OK** to enter the list of values to be tested. Confirm that the x-range is (-6.5 to 6.5) and draw the plots with **OK**. Here's the result:

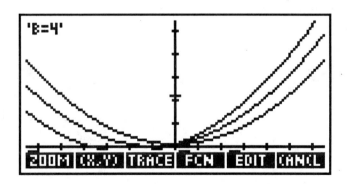

---

*Since you're going to evaluate a list of values, you must use the ADD function, not +. The + function *concatenates* lists (and *appends* or *prepends* objects to lists); the ADD function adds corresponding elements of lists—in a manner analogous to $\boxed{-}$, $\boxed{\times}$, and $\boxed{\div}$.

Sometimes a function appears in a less recognizable form, so that you need to do a little rearranging before it looks familiar.

Example: Symbolically rearrange $2x + 5 = \dfrac{(3 - y)}{x}$ so that y is isolated.

1. Use ⟶SYMBOLIC▼▼▼ ▇OK▇ to get the **ISOLATE A VARIABLE** screen. Note that this feature works only when the variable you want to isolate appears *exactly once* in the equation (so, for example, you could not use it to solve for x here).

2. Enter the equation into the **EXPR:** field: ⟵EQUATION 2 α⟵X +
5 ⟵= ⟵() 3 − α⟵Y ▶ ÷ α⟵X ENTER.

3. Enter **y** (note the lower-case) into the **VAR:** field: α⟵Y ENTER.

4. Make sure the **RESULT** will be **Symbolic** (if necessary, press +/− to toggle between **Numeric** and **Symbolic**). Press ▇OK▇ to let the application do the isolation for you.

   <u>Result:</u> `'y=3-(2*x+5)*x'`

5. Finally, press ⟵SYMBOLIC **EXPA COLCT** to tidy up the result:

   `'y=3-2*x^2-5*x'`

Now that the function is in a slightly more conventional form, you can analyze it in several different ways....

Example: Use *three different methods* to solve the result from the previous example for x: Graphic, numeric, and symbolic.

The Graphic Method:

1. Press ⇨(PLOT) to begin the **PLOT** application.

2. With the **EQ:** field highlighted, press (NXT) ▐ **CALC** ▌ then make sure the target equation is on stack level 1 (you may need to press (DROP)) and press ▐ **OK** ▌ to copy the equation into the **EQ:** field.

3. Make sure that **TYPE:** contains Function, that **INDEP:** contains (lowercase) ✕, that **H-VIEW:** is set to default values (−6.5 to 6.5), and that **AUTOSCALE** is checked.

4. Press ▐**ERASE**▐ ▐**DRAW**▐ to draw the plot.

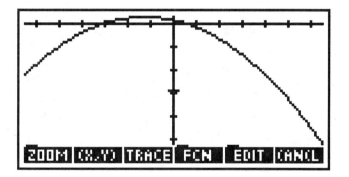

5. First, use the arrow keys (▲, ▼, etc.) to move the cursor to the parabola's apex. Then re-center the graph: Press ▐**ZOOM**▐ (NXT) ▐**CNTR**▐.

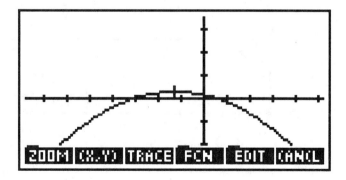

6. To zoom the graph at the region of interest (here it's where the graph crosses the *x*-axis), first move the cursor to the upper-left corner of the desired display area. Then press ZOOM BOX, then move the cursor so that the zoom-box is drawn over the region of interest; press ZOOM. The shape and location of your plot may differ slightly from the one shown below because the exact size and location of your zoom-box may differ slightly from the one used below.

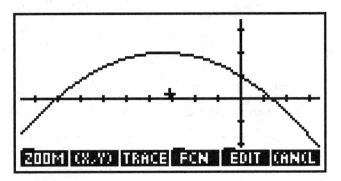

7. Press TRACE (X,Y). You can now use ◄ and ► to *trace along the graph*! Trace to the points where it meets the *x*-axis. Notice the cursor coordinates there. (Again, the coordinates you see may differ from the ones shown below because of differences in the zoom-boxes used.)

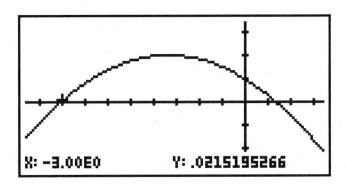

8. With the cursor near a solution point, press (NXT) (if the menu is hidden) FCN ROOT. Doing this for each root shows that the exact solutions are indeed -3 and 0.5.

The Numeric Approach

1. Return to the stack ([CANCEL][CANCEL]) then press [→][SOLVE] ▮**OK**▮ to begin the **SOLVE EQUATION** application.

2. If you just entered the equation in the **PLOT** application, you will probably see it displayed in the **EQ:** field. If not, enter it manually.

3. Enter a **0** value for **Y:** (the "solution" is the value of x when y is zero).

4. Enter a positive guess (say, 5) for **X:**. Then move the highlight back to the **X:** field and press ▮**SOLVE**▮. Result: **X: .5**

5. Try a negative guess (say, -5) for **X:**. Then, again, move the highlight back to the **X:** field and press ▮**SOLVE**▮. Result: **X: -3**

The Symbolic Approach

1. Return to the stack ([CANCEL]), then press [→][SYMBOLIC][▲][▲] ▮**OK**▮ to begin the **SOLVE QUADRATIC** application.

2. Enter the expression (**'3-2*x^2-5*x'**) into the **EXPR:** field: [←][EQUATION][3][−][2][α][←][X][yˣ][2][▶][−][5][α][←][X][ENTER].

3. Enter the variable for which you're solving (**X**) in the **VAR:** field.

4. With **RESULT:** showing **Symbolic** and **PRINCIPAL** unchecked, press ▮**OK**▮. The *general* solution results: **'x=(5+s1*7)/-4'**.

5. The **s1** in the solution is a placeholder variable inserted by the HP 48 that means "**±**." Thus, the single result is actually two results: the value when **s1** = 1 and the value when **s1** = -1. To further evaluate the two results, press [']['α][←][X] [←][PURGE][ENTER][1][']['α][←][S][1] [STO][EVAL][SWAP][1][+/−][VAR][←] ▮**s1**▮ [EVAL].

    Results: **'x=-3'** and **'x=.5'**

Rational Functions

A *rational* function is the quotient of two polynomial functions: $f(x) = \dfrac{p(x)}{q(x)}$
Wherever $q(x)$ is zero, the function is undefined; there exists a *discontinuity*. For example, the function $f(x) = \dfrac{2x}{x+5}$, has a discontinuity at $x = -5$. On the HP 48, whether (and how) such a discontinuity is displayed in a function plot depends on several factors:

- The *resolution* of the plot. The **PLOT** application plots only a sample of points along the function, so it simply may not sample the point of discontinuity. This may make it appear as if there is no discontinuity.

- The status of flag –31. When it is *set*, only the sampled points are plotted —no connecting line segments. Discontinuities can masquerade as unsampled points. When flag –31 is *clear*, **CONNECT** mode is enabled and the sampled points on either side of a discontinuity may be connected by a line segment—for many rational functions, a "vertical line" at the asymptote.

Example: Plot the rational function $f(x) = \dfrac{-3x}{x^2 + 3x + 2}$

1. Press →PLOT and make sure the **TYPE** is Function.

2. Enter the function in the **EQ:** field: ←EQUATION +/– 3 α ← X ÷ α ← X y^x 2 ▶ + 3 α ← X + 2 ENTER.

3. Set the **INDEP:** variable to (lowercase) X and set the **H-VIEW** range to −6 6 and the **V-VIEW** range to −2 5.

4. Bring up the **PLOT OPTIONS** screen (OPTS) and make sure that **CONNECT** is checked on (toggle the check-mark with either ✓CHK or +/–). Next, at the **STEP:** field, reset it to Dflt (default) value by pressing DEL ENTER. Then move the highlight to the bottom of the screen and enter 1 in the **H-TICK:** and **V-TICK:** fields and turn off the check-mark in the **PIXELS** field on the bottom line. This will place a tick-mark for each one unit along the horizontal and vertical axes —no matter what your display settings are.

5. Press OK ERASE DRAW to draw the plot.

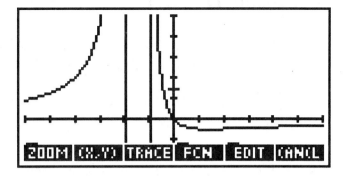

Note the vertical lines representing asymptotes near $x = -2$ and $x = -1$. The undefined points at the asymptotes lie *between* two plotted pixels, which were connected (in **CONNECT** mode) by an apparently vertical line segment.

Example: Plot this rational function:

$$f(x) = \frac{-3x}{x^2 - 3x + 2}$$

1. Press (CANCEL) to return to the **PLOT** input form and reset the plot parameters: (DEL)(▼)(ENTER).

2. Highlight the **EQ:** field, press **EDIT**, change the first **+** in the denominator of the function to a **−** (eleven (▶)'s, (◀), then (−)(ENTER)).

3. Change **INDEP:** back to **X** and the values in **Y-VIEW** to **−35** and **35**. Press **ERASE DRAW** to draw the plot.

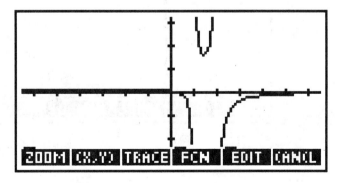

Note that no "vertical" asymptote lines appear because the function is undefined for the exact value of one of the pixels.

Exponential and Logarithmic Functions

Exponential functions have the form $f(x) = a^x$, where $a \neq 1$. The variable is contained in the *exponent* part of the expression.

Logarithmic functions are inverse functions of exponential functions. They have the form $f(x) = \log_a x$.

Example: Use FMPLT to explore the effect of varying the *a* parameter in an exponential function.

1. From the stack, type FMPLT and press (ENTER) or select FMPL from the (VAR) menu.

2. Enter the exponential formula (' a^x ') into the EQ: field and X into the INDEP: field (if it isn't already there).

3. Enter the parameter that you are varying (a) into the VARY: field.

4. Enter the list of values you want to use for a in the VALS: field. Try { 2 3 4 5 6 }.

5. Enter the horizontal range you want displayed in XMIN and XMAX. Use −1 to 2 .

6. Press OK .

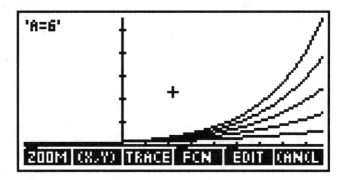

Example: Now use FMPLT to explore the effect of varying the *a* parameter in a logarithmic function.

Note that the HP 48 can use logarithms directly with only two bases —10 and *e*. However, a simple transformation makes the use of any logarithmic base possible:

$$\log_a x = \frac{\log_{10} x}{\log_{10} a}$$

1. Press (CANCEL) to return to the **FAMILY PLOT** input form.

2. Into the **EQ:** field, enter the transformed logarithmic formula,

    'LOG(x)/LOG(a)'

3. Leave the remaining entries as they were in the previous example and press **OK**.

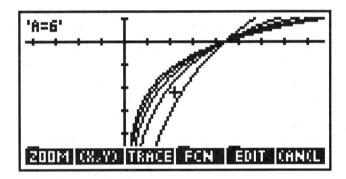

Compositions of Functions

If f and g are two functions, then the *composition* of g with f is the function $(g \circ f)(x) = g(f(x))$. Compositions use the output of one function as the input of the other. Not surprisingly, this makes the HP 48 very well-suited for performing compositions.

Example: Find the composition $g \circ f$ for

$$f(x) = \frac{2}{x+1} \text{ and } g(x) = \sqrt{4 - x^2}$$

1. Enter the f function ('2/(x+1)'); store it as 'f': ['][α][←][F][STO].
2. Enter the g expression, using f as its variable ('√(4-f^2)').
3. Press [3][+/−][α][α][C][F][ENTER] to be sure that you obtain symbolic results, then [EVAL]: '√(4-(2/(x+1))^2)'. If you want to look at the result in the EquationWriter, press [▼]. Then [CANCEL][CANCEL] to return to the stack when you've finished viewing it.

Example: Find the composition $f \circ g$ for the functions in the previous example, using the CMPOS program (see page 277).

1. Enter f: [←][EQUATION][2][÷][α][←][X][+][1][ENTER].
2. Enter g: [←][EQUATION][√x][4][−][α][←][X][yˣ][2][ENTER].
3. Enter the name of the variable in f: ['][α][←][X][ENTER].
4. Type CMPOS and press [ENTER] or press **CMPO** from the [VAR] menu.

Result: '2/(√(4-x^2)+1)'

Inverses of Functions

Two functions, $f(x)$ and $g(x)$, are *inverses* of each other if $f \circ g = x$ and $g \circ f = x$. For example, $2x$ and $\frac{1}{2}x$ are inverse functions of each other because both compositions yield x: $2(\frac{1}{2}x) = x$ and $\frac{1}{2}(2x) = x$.

Example: Find $f^{-1}(x)$, the inverse of $f(x) = \dfrac{3x - 1}{2}$.

1. Enter the function as an equation, substituting y for $f(x)$:

$$\texttt{'y=(3*x-1)/2'}$$

2. Solve for x by entering $\texttt{'x'}$ on the stack and pressing ←[SYMBOLIC] [ISOL]. The result is $\texttt{'x=(y*2+1)/3'}$. Note that ISOL works only when the solution variable ($\texttt{'x'}$ in this case) exists exactly once in the expression.

3. If you mentally exchange the positions of the x and y variables and substitute $f^{-1}(x)$ for y, you'll get the inverse function:

$$f^{-1}(x) = \frac{2x + 1}{3}$$

Example: The short program FINV (see page 281) makes it easier than that. To repeat the above example:

1. First, enter the function: ←[EQUATION] ←[()] [3] [α] ←[X] [−] [1] [▶] [÷] [2] [ENTER].

2. Then enter the variable of the function: ['] [α] ←[X] [ENTER].

3. Type in FINV and press [ENTER] or press [FINV] in the [VAR] menu.

   Note: Just like the built-in ISOL, the FINV program requires that the solution variable appears only once in the original expression.

   <u>Result:</u> $\texttt{'(x*2+1)/3'}$

User-Defined Functions

The HP 48 allows you to create short programs that work for the most part like the built-in mathematical functions. These programs, called *user-defined functions* (UDFs) all have the following structure:

$$\text{« → } \textit{local names} \quad \textit{defining procedure} \text{ »}$$

There should be one local name for each variable in the function you are defining. The defining procedure may either be a program (i.e. in postfix syntax) or an algebraic object.

The following examples illustrate a variety of user-defined functions.

Example: Create a UDF for computing the volume of a cone from the radius of its base and its height: $V = \frac{\pi}{3}r^2h$.

1. Type « → r h 'r^2*h/3*π' » (if you prefer to use the algebraic syntax in the defining procedure); or
   type « → r h « r SQ h * 3 / π * » » » (if you prefer the postfix syntax).
   Note that in either case, the π term comes last so that the remaining factors will be fully evaluated.

2. Enter the name for the UDF: ['] [α][α][V][C][O][N][E] [ENTER].

3. Store the function: [STO].

4. Test the function. Put a radius of 4 and a height of 11 onto the stack; execute the function VCONE: [4][ENTER][1][1][ENTER][VAR] VCON .

   Result: '58.6666666667*π' (if Flags -2 and -3 are clear)
   184.306769011 (if either Flag -2 or -3 is set).

   Reminder: Flag -2 controls whether symbolic constants such as π are kept symbolic (*clear*) or forced to be numeric (*set*). Flag -3 controls whether algebraic results are allowed to remain symbolic (*clear*) or are forced to be converted to numeric form (*set*).

Example: Create a user-defined function for the distance between two points in space:

$$\text{DIST}(x_1, y_1, z_1, x_2, y_2, z_2) = \sqrt{(x_2 - x_1)^2 + (y_2 - y_1)^2 + (z_2 - z_1)^2}$$

Assume that the function will find the six coordinates on the stack in the order shown next to DIST in the above definition.

1. Enter the equation, including the name, DIST, and the list of variables it uses (in the order they go on the stack) on the left-hand side:

[←][EQUATION][α][α][D][I][S][T][α][←][( )][α][←][X][1][←][,][α][←][Y]
[1][←][,][α][←][Z][1][←][,][α][←][X][2][←][,][α][←][Y][2][←][,]
[α][←][Z][2][▶][←][=][√x̄][←][( )][α][←][X][2][−][α][←][X][1][▶][yˣ][2][▶]
[+][←][( )][α][←][Y][2][−][α][←][Y][1][▶][yˣ][2][▶][+][←][( )][α][←][Z][2]
[−][α][←][Z][1][▶][yˣ][2][ENTER]

2. To store the expression on the right-hand side of the equal sign in the name on the left-hand side, you simply *define* the equation: [←][DEF].

3. Test it by finding the distance between the two points (3,-4,6) and (1,8,-3). Enter the six coordinates in order and execute DIST:

[3][ENTER][4][+/−][ENTER][6][ENTER][1][ENTER][8][ENTER][3][+/−][ENTER]
[VAR] DIST .

Result: 15.1327459504

Can a UDF be plotted? Usually not without modification; most UDF's remove objects from the stack (which wreaks havoc with the **PLOT** application). However, the modifications necessary to make them "plot-ready" are often quite easy, as the following example illustrates.

Example: Use the **VCONE** function to plot the variation of a cone's volume with the radius of its base (assuming the height remains constant).

1. Open the **PLOT** application and highlight the **EQ:** field.

2. Move to the stack: (NXT) **CALC**.

3. Recall **VCONE** to the stack and edit it so that r and h are given the values `'r'` and 1, respectively: (VAR)(→) **VCON** (▼)(▶)(')(α)(←)(R) (▶)(SPC)(1)(ENTER).

4. Return the modified version to the **EQ:** field: (←)(CONT) **OK** (NXT).

5. Put **r** into **INDEP:** and set **H-VIEW:** −.5 6 and **V-VIEW:** −5 20.

6. Press **OPTS** to move to the **PLOT OPTIONS** screen. On the top line you will see settings for the plotting range for the independent variable. Much of the time, the plotting range is the same as the horizontal display range (**H-VIEW**)—indeed, this is the default setting. However, there are occasions when you may wish to *plot* a different set of values than those indicated in your choice of *display* range. Set the plotting range to **LO: 0 HI: 5**. Then press **OK** to save your settings and return to the main PLOT screen.

7. Plot the function: **ERASE DRAW**:

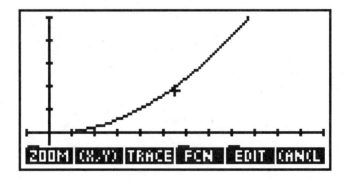

You may prefer to use a friendlier kind of user interface with your UDF's. Jim Donnelly's program, UDFUI, included in his book *The HP 48 Handbook* (and included here with his permission), takes the name of a UDF that you have already created and provides it the kind of friendly user-interface similar to many built-in applications on the G-Series machines.

Example: (This example assumes that you have already keyed in the UDFUI program, listed on page 317, and that it is stored in the current directory path). Use the UDFUI program to get a special interface for the UDF in the example on page 25 (DIST).

1. Enter the name of the UDF onto the stack: `'` `α` `α` `D` `I` `S` `T` `ENTER`.

2. Execute the UDFUI program: `α` `α` `U` `D` `F` `U` `I` `ENTER`.*

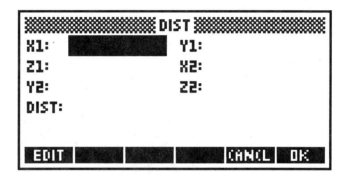

3. To test the new interface, calculate the distance between the points (4,-1,-2) and (-5,3,0). Move the highlight to the **X1:** field and enter the various coordinates: `4` `ENTER` `1` `+/−` `ENTER` `2` `+/−` `ENTER` `5` `+/−` `ENTER` `3` `ENTER` `0` `ENTER`; press **OK** , then `▲` **EDIT**:

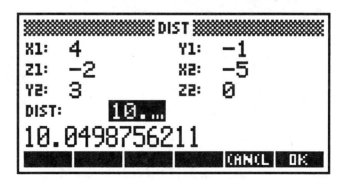

*If you have trouble here, be sure that DIST must be in the current directory path in order for UDFUI to find it.

2. TRIGONOMETRY

The Trigonometric Relationships

Trigonometry is the use of "triangle measurements" to describe *angles*. Every angle θ has an associated a set of right triangles that can be constructed around it to demonstrate various relationships:

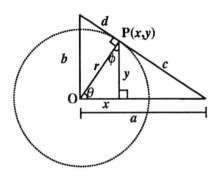

$$\sin \theta = \frac{y}{r} \qquad \cos \theta = \frac{x}{r} \qquad \tan \theta = \frac{y}{x} = \frac{c}{r}$$

$$\csc \theta = \frac{r}{y} = \frac{b}{r} \qquad \sec \theta = \frac{r}{x} = \frac{a}{r} \qquad \cot \theta = \frac{x}{y} = \frac{d}{r}$$

The six trigonometric relationships—*ratios,* actually—are derived directly from the basic geometric rules of similar triangles. And notice that if you let the radius $r = 1$ (i.e. use a *unit circle*), then the ratios show up even more directly:

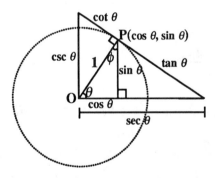

Besides showing the trigonometric ratios directly, the unit circle also helps to define trigonometric *functions*. These functions describe how the trigonometric ratios (sine, tangent, etc.) *change* as the angle θ changes (e.g. as the radius, r, "sweeps" around the circle). The trigonometric ratios encountered as the radius sweeps around the circle repeat themselves; i.e., the trig functions are *periodic*.

Radians and Degrees: Units of Angle

When measuring angles as a part of triangles, it is most common to use *degrees*, *minutes* and *seconds*. But whenever you want to use angle measure as the independent variable in a function—for example, when plotting or solving—you should use units of *radians*. A radian is the amount of angle you sweep out as you move along the arc of a circle for a distance equal to the radius of that circle. One radian is exactly $\frac{180}{\pi}$ degrees (approximately 57.3°).

Example: Convert 214° to radians.

1. Enter 214 onto the stack: [2][1][4][ENTER].
2. Press [MTH] **REAL** [NXT][NXT] **D→R**.
   Result: **3.73500459927** radians

Example: Add 23° 34' 18" to 15° 42' 07" and convert the result to radians.

1. Enter 23.3418 onto the stack: [2][3][·][3][4][1][8][ENTER].
2. Enter 15.4207 and press [←][TIME][NXT] **HMS+** (HMS+ does degree *and* time addition, hence the acronym: "Hours, Minutes, Seconds").
3. Convert the result to *decimal degrees*: **HMS→**.
4. Convert decimal degrees to radians: [MTH] **REAL** [NXT][NXT] **D→R**.
   Result: **.685453823037**

Example: Convert $\frac{\pi}{28}$ radians to degrees, minutes and seconds.

1. Enter $'\pi/28'$ onto the stack: ['][←][π][÷][2][8][ENTER].
2. First, convert this angle measure to decimal degrees: [MTH] **REAL** [NXT][NXT] **R→D** (then [←][→NUM] if necessary).
3. Now convert from decimal degrees to degrees, minutes, and seconds: [←][TIME][NXT]**→HMS**.
   Result: $'6.25428571428'$, which means 6° 25' 42$\frac{6}{7}$"

Example: Plot the function, sin(X), in radians, then in degrees.

1. Begin the **PLOT** application ((→)(PLOT)) and make sure that the **TYPE** is **Function** and that **∡** is **Rad**.

2. Reset the plot parameters to their defaults: (DEL)(▼)(ENTER).

3. With the highlight on the **EQ** field, type (')(SIN)(α)(X)(ENTER).

4. Draw the plot: **ERASE DRAW**.

5. When the plot is drawn, press (CANCEL) to return to the **PLOT** screen, and change **∡** to **Deg** (press (←)(RAD)).

6. Draw the plot again on top of the previous one: **DRAW**:

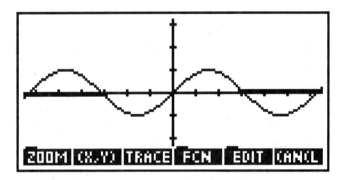

Those two plots are not nearly the same, are they? This illustrates the importance of matching the display ranges to the angle measure: To achieve the same plot in degrees, you would need to change the **H-VIEW** to run from -720 to 720 before drawing.

The TRIGX Program

TRIGX is a program that computes a number of different ratios and values for a given angle. After entering the program (see page 314 in the Appendix), you start it running either by typing TRIGX and pressing (ENTER) or selecting **TRIGX** from the (VAR) menu.

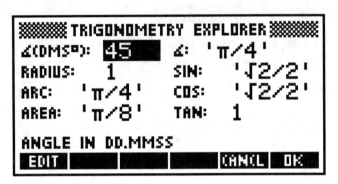

To use TRIGX, you simply enter any known values into their appropriate fields and then press **OK**. The program will compute values for the remaining fields. Here's how it works:

- If the angle measure and radius are given, TRIGX bases its computations on them.

- If more than one angle measure is given (i.e. degrees and radians), then the computations are based on the currently set mode; the other is re-computed to match.

- If there is no angle measure or radius given, they are computed from those values that *are* given, if at all possible. Only principal angles (matching the given inputs) will be returned.

- If it is not possible to compute an angle measure or a radius, default values of 45° and 1 are used and the computations adjusted.

Example: Given: 60° angle and a radius of 3.

   1. Begin the program and enter the values in the appropriate fields.

   2. Press ▓ OK ▓ .

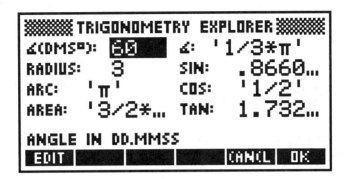

Example: Given: $\cos \theta = -0.5$, $\tan \theta = \sqrt{3}$, and arc length $= \dfrac{40\pi}{3}$.

   1. Clear the data from the previous example: DEL ▼ ENTER.

   2. Enter values in the appropriate fields. Note that you can use tick-marks to enter a value "symbolically" (i.e. '40/3*π'), if you wish:
   ▼ ▼ ' 4 0 ÷ 3 × ← π ENTER . 5 +/− ENTER ▶ ' √x 3
   ENTER.

   3. Press ▓ OK ▓ .

Verifying Identities

There are a number of special interrelationships between the trigonometric functions that are always true no matter the size of the angle or angles involved. These interrelationships are called *identities*.

Here is a core list of important trigonometric identities:

- **Pythagorean Identities.** These are evident if you apply the Pythagorean theorem for right triangles to the diagram at the bottom of page 29.

$$\sin^2\theta + \cos^2\theta \equiv 1 \quad (\text{thus } 1 - \sin^2\theta \equiv \cos^2\theta \text{ and } 1 - \cos^2\theta \equiv \sin^2\theta)$$

$$1 + \tan^2\theta \equiv \sec^2\theta \quad (\text{thus } \sec^2\theta - \tan^2\theta \equiv 1 \text{ and } \sec^2\theta - 1 \equiv \tan^2\theta)$$

$$1 + \cot^2\theta \equiv \csc^2\theta \quad (\text{thus } \csc^2\theta - \cot^2\theta \equiv 1 \text{ and } \csc^2\theta - 1 \equiv \cot^2\theta)$$

- **Difference and Sum Identities.**

$$\cos(\alpha \pm \beta) \equiv \cos\alpha\cos\beta \mp \sin\alpha\sin\beta$$

$$\sin(\alpha \pm \beta) \equiv \sin\alpha\cos\beta \pm \cos\alpha\sin\beta$$

$$\tan(\alpha \pm \beta) \equiv \frac{\tan\alpha \pm \tan\beta}{1 \mp \tan\alpha\tan\beta}$$

- **Double Angle Identities.**

$$\sin 2\theta \equiv 2\sin\theta\cos\theta$$

$$\cos 2\theta \equiv \cos^2\theta - \sin^2\theta$$
$$\equiv 1 - 2\sin^2\theta$$
$$\equiv 2\cos^2\theta - 1$$

$$\tan 2\theta \equiv \frac{2\tan\theta}{1 - \tan^2\theta}$$

- **Half-Angle Identities.**

$$\sin\frac{\alpha}{2} \equiv \pm\sqrt{\frac{1-\cos\alpha}{2}} \qquad \cos\frac{\alpha}{2} \equiv \pm\sqrt{\frac{1+\cos\alpha}{2}} \qquad \tan\frac{\alpha}{2} \equiv \pm\sqrt{\frac{1-\cos\alpha}{1+\cos\alpha}}$$

- **Sum and Product Identities.**

$$\sin\alpha + \sin\beta = 2\sin\frac{\alpha+\beta}{2}\cos\frac{\alpha-\beta}{2}$$

$$\sin\alpha - \sin\beta = 2\cos\frac{\alpha+\beta}{2}\sin\frac{\alpha-\beta}{2}$$

$$\cos\alpha + \cos\beta = 2\cos\frac{\alpha+\beta}{2}\cos\frac{\alpha-\beta}{2}$$

$$\cos\alpha - \cos\beta = -2\sin\frac{\alpha+\beta}{2}\sin\frac{\alpha-\beta}{2}$$

These core identities, the proofs for which are usually included in standard math textbooks, are themselves used in two important ways:

- To establish new identities that are useful in particular problem-solving situations.
- To aid in obtaining exact numerical solutions to problems involving trigonometric solutions.

Look at each of these uses, one at a time....

To use the HP 48 to help establish new trigonometric identities, you need to use its symbolic manipulation tools.

When doing symbolic manipulations, be sure that flag –3 is clear (press →MODES FLAG and make sure that flag 03 is clear—i.e. unchecked) so that results remain symbolic.

Example: Verify that $\dfrac{1 - \cos x}{\sin x} = \dfrac{\sin x}{1 + \cos x}$ is indeed an identity.

Although it is often faster and more convenient to do this kind of algebraic manipulation manually, the HP 48 is capable of performing symbolic verifications.

1. Type the expression in the EquationWriter, ←EQUATION ▲ 1 ⊟ COS α ←X ▶ ▶ SIN α ←X ▶ ▶ ←⊟ SIN α ←X ▶ ÷ 1 ⊞ COS α ←X, and press ENTER to put it onto the stack.

2. Multiply both sides of the equation by sin x: ' SIN α ←X ENTER ⊠.

3. Simplify the result by collecting like terms: ←SYMBOLIC COLCT.

4. Multiply both sides of the equation by $1 + \cos x$ and simplify the results: ' 1 ⊞ COS α ←X ENTER ⊠ COLCT.

5. Use the pattern matching application to replace 'SIN(x)^2' with its equivalent, '1−COS(x)^2': First, press →SYMBOLIC ▲ OK MATC to display the MATCH EXPRESSION screen.

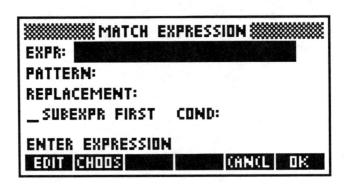

Next, with the **EXPR:** field highlighted, retrieve the expression: Press (NXT) **CALC** (and ◄ if the target expression isn't already on level 1), then **OK** .

Now highlight the **PATTERN** field, and enter the pattern to be replaced, substituting each occurrence of **x** with the wildcard name **&1**: (')(SIN)(α)(◄)(ENTER)(1)(►)(yˣ)(2)(ENTER). (Notice that the special wildcard character, **&**, can be typed with (α)(◄)(ENTER).)

Type in the replacement pattern, using the wildcard name instead of the variable: (')(1)(−)(COS)(α)(◄)(ENTER)(1)(►)(yˣ)(2)(ENTER). Press (ENTER)(ENTER) to return the modified expression to the stack. Press **EXPA** **EXPA** **EXPA** **COLCT**.

<u>Result</u>: `'1-COS(x)^2=1-COS(x)^2'`

The verification is complete.

Example: Verify that $\sin^4 x + 1 = 2\sin^2 x + \cos^4 x$.

This time use a graphical approach. Notice that if you subtract $\sin^4 x$ from both sides of the target equation, you will have an expression equal to 1, so it should (if the original equation is true) produce a horizontal line plot.

1. Enter the expression onto level 1 of the stack:

   $$\texttt{'SIN(x)\^{}4+1=2*SIN(x)\^{}2+COS(x)\^{}4'}$$

2. Subtract $\sin^4 x$ from both sides of the equation: [']$\,$[SIN]$\,$[α]$\,$[←]$\,$[X]$\,$[▶] [y^x][4][ENTER][−]⬛COLCT.

3. Press [→][PLOT] to prepare to plot the equation.

4. Reset the plot parameters to their defaults: [DEL][▼][ENTER].

5. Make sure that **TYPE** is **Function** and ∡ is **Rad**. Then, with the **EQ** field highlighted, press [NXT]⬛CALC ([←], if necessary to bring the target equation to level 1) [←][EDIT][▶][DEL][DEL][ENTER] ⬛OK⬛ [NXT]. This puts the right-hand side expression into **EQ**.

6. Enter **x** (lower-case) in **INDEP** and press ⬛ERASE⬛DRAW⬛:

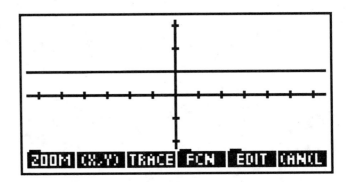

7. Press ⬛TRACE⬛(X,Y)⬛ and then move along the plot with [◀] and [▶] to convince yourself that the expression equals 1 for all values of **x**.

8. Although this seems to confirm the identity, it isn't *proof*. Try one more thing before you accept the verdict for good. Press (CANCEL) to return to **PLOT** screen, check **AUTOSCALE** and press **DRAW** to superimpose the autoscaled plot on top of the original:

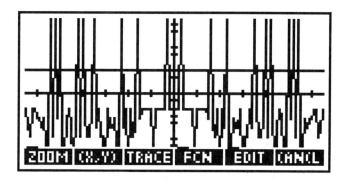

What's happening?!? Why did the seemingly constant graph suddenly become very "non-constant?"

To find out, press (CANCEL) and inspect the **Y-VIEW** parameters that were automatically computed by the machine (highlight them and press **EDIT**).... You will discover that the variation you are viewing is occurring in a vertical range of 0.000000000016. Thus, you can conclude that the identity is true: the "huge" visual variation you see is actually negligible—caused only by the 12-digit limitation on numerical precision in the HP 48.

So although it can't provide rigorous *proof*, the graphical approach to identity verification is a good check against your symbolic derivations (whether performed by hand or on the HP 48). However, keep in mind that autoscaling tends to show you *any* variation it finds, and thus it can fall prey to the machine's numerical limitations (round-off), as in this case.

The next example illustrates the computational use of identities.

Example: Solve this equation for x: $\sin\dfrac{17\pi}{12} - \sin\dfrac{11\pi}{12} = x\sin\dfrac{\pi}{4}$

You want an *exact* answer, not merely a 12-digit numerical approximation, so be sure that flag −3 is clear before trying to work symbolically.

1. Notice that the left-hand side of the equation matches the form of one of the Sum and Product Identities (see page 35). So, make a pattern substitution:

   First, from the stack, press ⏵(SYMBOLIC)(▲) ▮OK▮ ▮MATC▮. Then enter the equation in the **EXPR** field: ⏴(EQUATION)(SIN)(1)(7)⏴(π)(÷) (1)(2)(▶)(▶)(−)(SIN)(1)(1)⏴(π)(÷)(1)(2)(▶)(▶)⏴(=)(α)⏴(X)(SIN)⏴(π) (÷)(4)(ENTER).

   Next, enter the pattern to be matched in **PATTERN**:
   $$\texttt{'SIN(\&1)-SIN(\&2)'}$$

   And enter the new pattern in **REPLACEMENT**:
   $$\texttt{'2*COS((\&1+\&2)/2)*SIN((\&1-\&2)/2)'}$$

   Press (ENTER)(ENTER)(ENTER) to make the replacement and return to the stack.

2. Solve for X: (')(α)⏴(X)⏴(SYMBOLIC) ▮ISOL▮▮COLCT▮

3. Change the decimals to fractions. Note that whenever you do this, STD display format sometimes yields odd results do to rounded precision in the last decimal place (thus a change to FIX format): (9) (SPC)(α)(α)(F)(I)(X)(ENTER)(NXT) ▮→Q▮.

4. Collect terms again and convert the decimals in the resulting equation to fractions once again: ⏴(PREV)▮COLCT▮(NXT) ▮→Q▮.
   $$\underline{\text{Result:}} \quad \texttt{'x=2*COS(7/6*π)'}$$

Varying Coefficients in Trig Functions

The sine and cosine functions can be described generically as:

$$f(x) = A\sin^E(Bx + C) + D \qquad \text{or} \qquad f(x) = A\cos^E(Bx + C) + D$$

Several characteristics of these plots of this function can be determined by its coefficients:

- The *amplitude* of the function—the height (or depth) of each "wave"— is equal to $|A|$.

- The *period* of the function is equal to $\dfrac{2\pi}{|B|}$.

- The *horizontal* (or *phase*) *shift* of the function is $-\dfrac{C}{|B|}$.

- The *vertical shift* of the function is D.

- The *shape* of the curve is affected by E. Higher values of E yield steeper, more jagged curves.

You can investigate all of these characteristics more thoroughly by using the FMPLT program and the generic function shown above. Here are some sample results, shown in the next few figures....

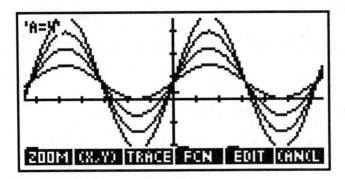

Different Amplitudes
A={1 2 3 4}, B=1, C=0, D=1, E=1

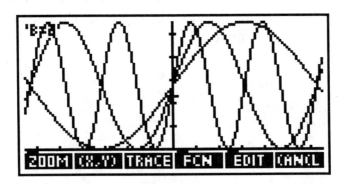

Different Periods
A=1, B={.5 1 2}, C=0, D=1, E=1

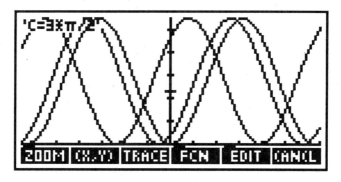

Horizontal Shifting
A=1, B=1, C={'-π/3' 'π/4' '3*π/2'}, D=1, E=1

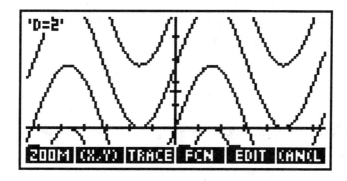

Vertical Shifting
A=1, B=1, C=0, D={-1 0 1 2}, E=1

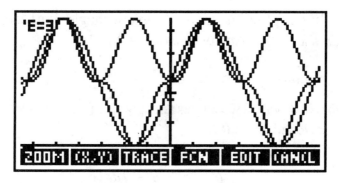

Different Curve Shape
A=1, B=1, C=0, D=1, E={1 2 3}

Varying Coefficients in Trig Functions   43

Solving Triangles

One of the most important uses for trigonometry is the computation of distances or angles that cannot be measured directly. Here's the general approach for these kinds of problems:

- "Create" a triangle involving the unknown measurement as one of its sides or angles. This is sometimes called *triangulation*.

- Measure two or more *accessible* elements (sides and angles) of the triangle.

- Use the principals and theorems of trigonometry to compute the unknown, remote element.

The process of computing the missing elements of a triangle from a few givens is referred to as *solving the triangle*. The figure below shows the elements of the triangle as they are conventionally labeled:

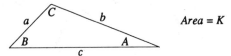

Area = K

Note that *a* is the shortest side, *c* the longest. Angle *A* is opposite side *a*, angle *B* opposite side *b*, and angle *C* opposite side *c*.

Triangle solutions use of a series of trigonometric laws, each of which requires that a particular set of triangle elements be known:

- **Sum of the angles**: The sum of the interior angles in a triangle equals 180°. *Required knowns:* Any two angles (AA).

- **Law of Sines:** $$\frac{a}{\sin A} = \frac{b}{\sin B} = \frac{c}{\sin C}$$
Required knowns: Two angles and any side (AAS, ASA); or two sides and a non-included angle (SSA).

- **Law of Cosines:** $$a^2 = b^2 + c^2 - 2bc\cos A$$
$$b^2 = a^2 + c^2 - 2ac\cos B$$
$$c^2 = a^2 + b^2 - 2ab\cos C$$
Required knowns: Two sides and the included angle (SAS); or all three sides (SSS).

- **Heron's Formula:** $K = \sqrt{s(s-a)(s-b)(s-c)}$, where $s = \dfrac{a+b+c}{2}$

  Required knowns: All sides (SSS); or the area and any two sides (KSS).

- **Area Formula:** $K = \frac{1}{2}ab\sin C = \frac{1}{2}ac\sin B = \frac{1}{2}bc\sin A$

  Required knowns: Two sides and the included angle (SAS); or the area and any two sides (KSS); or the area, an angle and an adjacent side (KSA).

- **Area Formula (2-angle form):** $K = \dfrac{1}{2}a^2\,\dfrac{\sin B\sin C}{\sin(B+C)}$

  Required knowns: Two angles and the included side (ASA); or the area and any two angles (KAA); or the area and an angle and an adjacent side (KSA).

Example: Solve a triangle (including its area), given: $A = 25°$, $b = 6$, $c = 3$

1. Consider which laws you can apply to obtain the missing elements. The known values here are two sides and the included angle (SAS). Thus the Law of Cosines and Area Formula will be useful first steps.

2. Solve for *a* using the Law of Cosines: Make sure that you're in **DEG** mode and then press ⏵SOLVE ▮OK▮ to begin the Solver. With the highlight on the **EQ** field, press ' α⏴A y^x 2 ⏴ = α⏴B y^x 2 + α⏴C y^x 2 − 2 × α⏴B × α⏴C × COS α A α ⏵ 6 ENTER. Note that we are using a degree mark following the letter for angle names. Enter values in the **A°**, **B**, and **C** fields. Highlight the **A** field and press ▮SOLVE▮.

   Result: <u>**A: 3.51751612174**</u>

3. Use the Law of Sines to compute *B*: Move the highlight to **EQ** and enter the appropriate equation: ' a∕SIN(A°)=b∕SIN(B°) '. The known values are still there from before, so highlight the **B°** field and press ▮SOLVE▮. <u>Result: **B°: 46.1272426359**</u>

4. Subtract the sum of **A°** and **B°** from 180° to find **C°**: Press CANCEL 2 5 + 1 8 0 ⏴SWAP −. <u>Result: **108.872757364**</u>

5. Finally, compute $K = \frac{1}{2}bc\sin A$: Press 2 5 SIN 6 × 3 × 2 ÷.

   <u>Result: **3.80356435568**</u>

Example: Solve a triangle (including its area), given: $A = 15°$, $a = 4$, $b = 8$.

This time, use the program SOL⌂ (for the program listing, see page 308 in the Appendix) to automate the process of solving for multiple missing variables.

1. Type SOL⌂ and press (ENTER); or select **SOL⌂** from the (VAR) menu.

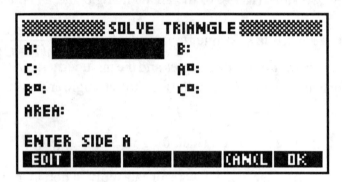

2. Enter the known values into their appropriate fields: (4)(ENTER)(8) (ENTER)(▶)(1)(5)(ENTER).

3. Press **OK**. After a moment, you will get a message indicating that the program has found **One of two solutions**.

This indicates that there are actually two different triangles that can have $A° = 15°$, $a = 4$, and $b = 8$. Press **OK** and one of them will be returned to the **SOLVE TRIANGLE** screen:

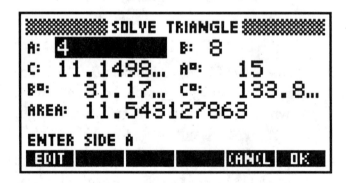

4. The two solutions use different *supplementary* values for $B°$, which was the "missing" member of the two "couples"—*a* and $A°$, *b* and $B°$—in the original problem. So, find the supplement of $B°$, using the stack, then delete the values for c, $C°$ and the area, and compute the other solution: ▼▼ NXT CALC 1 8 0 ENTER SWAP ─ OK ▲ DEL ENTER ▼ ▶ DEL ENTER ▼ DEL ENTER OK .

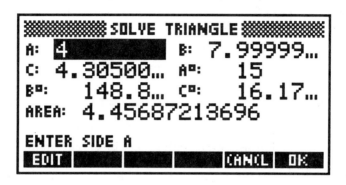

Example: Solve a triangle given: $K = 25, A = 38°, C = 86°$

1. Press DEL ▼ ENTER to reset the values in the **SOLVE TRIANGLE** screen.

2. Enter the known values into their appropriate fields: ▼ ▶ 3 8 ENTER ▶ 8 6 ENTER 2 5 ENTER.

3. Press OK .

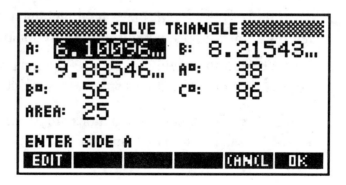

Solving Trigonometric Equations

Recall that an equation that is true for *all* values of the independent variable is called an *identity*. However, many useful problems and real-world situations can be modeled using conditional equations—which are only true for a small subset of values of the independent variable.

For example, the equation $\sin \theta = \frac{1}{2}$ is a conditional equation because it is true for only *some* values of θ. To determine these particular values, the equation must be *solved*. There are three different approaches:

- Use the built-in root-finder, with either the Solver or the Function Plot Analysis tools.

- Compute solutions directly from the keyboard using the inverse trigonometric functions.

- Use the ISOL command to symbolically isolate ("solve for") a particular variable.

Example: *Root-Finder.* Use the built-in root-finder in the SOLVE application to solve the equation $\sin \theta = \frac{1}{2}$.

1. Make sure that you're in Degree mode, press ⟨→⟩⟨SOLVE⟩ ▊OK▊ and enter the equation ('SIN(θ)=.5') in the EQ field (note that ⟨α⟩⟨→⟩⟨F⟩ types a θ).

2. Then press ▊SOLVE▊ with the θ field highlighted. The root-finder returns a solution, 30, but it isn't the only solution.

3. Enter 200 into the θ field as a guess, re-highlight that field, and press ▊SOLVE▊. Result: 150. Or try using a guess of 2000. Result: 1950.

    Remember: Trigonometric functions are *periodic* functions; they repeat the same values over and over as the independent variable increases or decreases.

Example: *Function Plot Analysis.* Plot the expression `'SIN(θ)-.5'` using ⟶ PLOT and solve for θ from the plot.

1. While viewing the **PLOT** screen, enter the expression into the **EQ** field. Change **INDEP** to θ and set the **H-VIEW** to **-2000 2000** and **V-VIEW** to **-2 1**.

2. Use the **PLOT OPTIONS** screen (press **OPTS**) to set the **STEP:** to **10**; press **OK**.

3. Finally, make sure that ∡ is set to **Deg** and press **ERASE DRAW**.

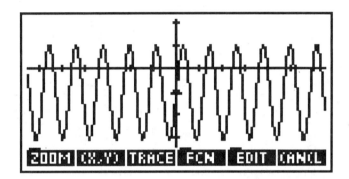

4. Each point where the plot crosses the horizontal axis is a solution. To find one, move the cursor out toward the right side of the screen and press **FCN ROOT**.

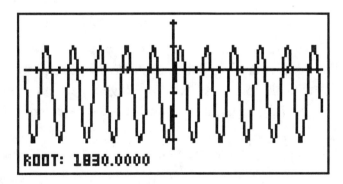

As with the **SOLVE** application, you may repeat this with any of the possible solutions.

Example: *Keyboard functions*. Find the $\sin \theta = \frac{1}{2}$ directly from the keyboard.

1. Enter .5 onto the stack.

2. Press ⬅|ASIN|. Result: 30.

The inverse trigonometric functions located on the keyboard (ASIN, ACOS, and ATAN) always return the *principal value* solution. For ASIN and ATAN, the principal value solution is that located between −90° and 90°. For ACOS, the principal value solution is that located between 0° and 180°.

Example: *The ISOL command*. Use the ISOL command to find the *general* solution for the equation $\sin \theta = \frac{1}{2}$.

1. Make sure that flag −1 is *clear*. The ISOL command (among others) will return the general solution if flag −1 is clear and the principal solution if flag −1 is set. Press ➡|MODES| **FLAG** and be sure that 01 General solutions is unchecked (if necessary, press **✓CHK**). Press |ENTER||ENTER| when finished.

2. Enter the equation onto the stack.

3. Enter the name of the variable for which you are solving ('θ').

4. Press ⬅|SYMBOLIC| **ISOL**.

$$\underline{\text{Result}}: \text{'}\theta=30*(-1)^\wedge n1+180*n1\text{'}$$

The n1 variable stands for any whole number. If you store a whole number in n1 and then evaluate the general solution (after purging the solution variable, θ) you'll get a particular solution. If you store 0 in n1 and evaluate, you'll get the *principal* solution.

For example: |'||α||➡||F||⬅||PURG| then,
|ENTER||2||'||α||⬅||N||1||STO||EVAL| yields: 'θ=390'
|DROP||ENTER||5||'||α||⬅||N||1||STO||EVAL| yields: 'θ=870'
|DROP||3||+/−||'||α||⬅||N||1||STO||EVAL| yields: 'θ=−570'

Problem Solving with Trigonometry

This section works through a set of typical problem situations where trigonometry is useful. For most of these examples, you will probably want to set the display mode to something suitable for real-world situations: [2][α][α][F][I][X][ENTER].

Prob. 1: You are considering buying a piece of land for which the asking price is $75,000. It is triangular plot of land has two sides which have length 400 feet and 600 feet. The angle between these sides is 46°20'. If comparable land is selling for $1.15 per square foot, should you pay the asking price?

1. To compute the area of the triangular plot, use the area formula ($K = \frac{1}{2}ab\sin C$) directly (make sure that you're in DEG mode):

   Enter the two sides: [4][0][0][→][UNITS] **LENG** **FT** [6][0][0] **FT**.
   Multiply the two side lengths together and then halve the product: [X][.][5][X]. Next, enter the angle in DD.MMSS form (46.2) and convert it to decimal degrees: [←][TIME][NXT] **HMS→**. Now find the sine and multiply it by the previous product to compute the area of the plot: [SIN][X]

   <u>Result:</u> **86804.28_ft^2**

2. Compute the market value of the plot of land: [1][.][1][5] [→][_][α]
   [←][4][/][→][UNITS] **AREA** **FT^2** [ENTER][X].

   <u>Result:</u> **99824.92_$**

Prob. 2: Two ranger stations located 10 km apart receive a distress call from a camper. Electronic equipment allows them to determine that the camper is at an angle of 71° from the first station and 100° from the second, each angle having as one side the line between the stations. Which station is closer to the camper? How far away is it?

1. This problem of "triangulation" is one involving two angles and the included side (ASA). However, simply drawing a fairly accurate picture will give you an answer to the first question:

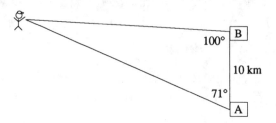

Obviously, the camper is closer to the station B, at the 100° angle.

2. To determine the distance, first find the third angle, using the law of the sum of the interior angles: $180° - (100° + 71°) = 9°$.

Now use the Law of Sines: $\dfrac{10 \text{ km}}{\sin 9°} = \dfrac{x \text{ km}}{\sin 71°}$

Press `1` `0` `→` `UNITS` `LENG` `NXT` `KM` `7` `1` `SIN` `×` `9` `SIN` `÷`.

Result: `60.44_km`

Prob. 3: Two tracking stations, located 1115 miles apart, simultaneously spot a UFO. One station measures the angle of elevation to be 28°, and the other 67° (relative to the same direction). How far above the surface of the earth is the UFO? How far away is it from the closest tracking station?

1. Draw a diagram of the problem:

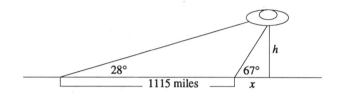

2. Note from the diagram that:

$$\tan 28° = \frac{h}{1115 + x} \qquad \text{and} \qquad \tan 67° = \frac{h}{x}$$

Thus, $h = (1115 + x)\tan 28° = x\tan 67°$

3. Be sure that you're in **DEG** mode, then input the equation and solve for x: 🢀EQUATION🢀() 1 1 1 5 + α🢀X ▶ TAN 2 8 ▶ 🢀= α🢀X TAN 6 7 ENTER ' α🢀X 🢀SYMBOLIC QUAD.

Result: '𝖷=325.01'

Prob. 4: Scientists at two astronomical observatories, located on the equator, observe the sun at the same time in order to determine its distance from earth. The observer at Observatory A, located at 135° 28' 13" West Longitude, views the sun exactly overhead. Meanwhile, her colleague at Observatory B, located at 45° 28' 22" West Longitude, simultaneously sees the sun centered on the horizon. If you assume that the earth's radius is 4000 miles, how far away is the sun?

1. Begin with a diagram. Use a view of the earth and sun from above the North Pole (not drawn to scale):

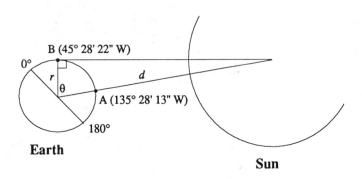

2. From the diagram, $d = \dfrac{r}{\cos\theta}$. The earth's radius, r, is 4000 miles, and the angle θ is the difference between the longitude of the two observatories. So, compute that angle: [α][α][S][T][D][ENTER][1][3][5][.] [2][8][1][3][ENTER][4][5][.][2][8][2][2][←][TIME][NXT] [HMS−].
Result: 89° 59' 51"

3. Now compute the distance d (and note that you must convert the angle to decimal form before finding the cosine): [HMS→][COS] [4][0][0][0][÷][1/x]. Result: 91673247.25 (miles)

Notes

3. POLAR AND PARAMETRIC EQUATIONS

The Parametric Perspective

The standard representation of a function, such as $f(x) = x^2$, implies that the function's output value *depends* upon the input value. To plot a point on the graph of a standard function you need only two things: the input value and the function expression. The horizontal coordinate (x-value) is *known*; only the vertical coordinate (y-value) need be computed.

Such "dependence" can be misleading, however, so functions may instead be represented *parametrically*. The parametric representation of a function requires that both the horizontal and vertical coordinates be computed via a *third* value—a *parameter*. The two most common parameters used are time (t) and angle (θ).

Obviously, the added complexity of parametric description must yield important additional value or no one would bother with it. For example, this function describes the curve a rock takes as it is thrown horizontally off a cliff at 32 ft/sec.:
$y = -x^2/64$.   That is, when the rock is a horizontal x feet from the cliff, it is y feet below its starting point. While that covers the raw facts of the observed motion, it doesn't help explain *why* the motion or make predictions about it, so it's hard to answer common questions: How long will it take the rock to land? Does throwing it faster forward make it hit the ground sooner? Where will it hit?

However, the parametric representation of this motion gives more information:

$$x = 32t$$
$$y = -16t^2$$

where t is the time (in seconds) after the throw.

Now you can see that the horizontal motion is unchanged from the initial throw, but that all of the acceleration is vertically downward due to gravity. And the inclusion of time into the function adds to the predictions you can then make.

Parametric representations are thus essential in separating and predicting the various components of complex motion. Parametric representations are compatible with the time-saving vector and matrix techniques of calculus and can, with no more complexity, be extended to any number of dimensions. Parametric methods are far easier to use in computer algorithms of all kinds—from the design of video-game images to the analysis of exploding particles in advanced physics.

Polar Coordinates

A point in a plane can be described in two distinct ways. **Rectangular** coordinates identify a point, P, by giving its horizontal and vertical position on a rectangular grid—(x,y). **Polar** coordinates identify that point, P, by giving its distance from the origin (or pole) and its direction with respect to the polar axis—(r,θ).

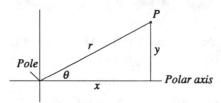

Functions using rectangular coordinates use the horizontal coordinate (usually x) as the independent variable. Functions using polar coordinates use the polar angle (usually θ) as the independent variable.

The relationship between rectangular and polar coordinates is best described as a special kind of parametric relationship—where the coordinates in one system are the parameters in the other:

$$x = r \cos \theta \qquad \text{or} \qquad r = \sqrt{x^2 + y^2}$$
$$y = r \sin \theta \qquad\qquad \theta = \tan^{-1}\left(\frac{y}{x}\right)$$

The HP 48 always treats the rectangular coordinate system as standard; it uses that representation internally when doing computations. However, it can display coordinates in either polar or rectangular mode, and you may enter coordinates in either form at any time. This means that you need to be careful! It is easy to confuse yourself and generate incorrect answers.

Look at some examples....

Example: In DEGree mode and with the rectangular coordinate $(3, 4)$ on the stack, change the display to Polar mode by pressing →POLAR (one of the polar annunciators is displayed).

Result: $(5, \angle 53.1301023542)$

The coordinates are *displayed* in polar form even though they are still stored internally in rectangular.

Example: While still in polar mode, enter the rectangular coordinate $(3, 4)$: ←()3 SPC 4 ENTER.

Result: $(5, \angle 53.1301023542)$

Although the coordinate displays in the command line as rectangular it's displayed in the current mode (polar) on the stack.

Example: Change the mode back to rectangular: →POLAR. Now enter the polar coordinate, $r = 5$ and $\theta = 60°$: ←()5 SPC →∠60 ENTER.

Result: $(2.5, 4.33012701892)$

Note that you must use the angle key to indicate that the second coordinate is an angle (i.e. that the entry is polar). And once again, the form displayed on the stack is that determined by the current setting (rectangular).

Example 4: Enter the polar coordinate $(5, \pi/3)$. Watch out! In polar mode, you must also pay attention to the angle mode (radians or degrees). Also, note that you cannot directly enter π into the coordinate. Press ←π ENTER 3 ÷ ←→NUM to compute $\pi/3$. Then ←RAD →POLAR ▼ ←()DEL 5 ←, →∠ ENTER.

Result: $(5, \angle 1.0471975512)$

Polar Representations and Complex Numbers

Complex numbers, such as $x + yi$, are comprised of two parts (real and imaginary). And they, too, have both rectangular and polar representations. Just as real numbers are plotted on a line, complex numbers are plotted on the complex plane—and thus are directly analogous to the coordinates of points in any plane.

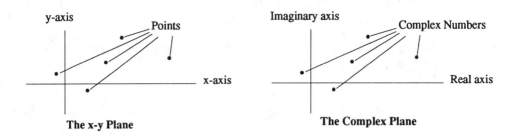

On the HP 48, complex numbers are typically represented by ordered pairs—just like the coordinates of points. Computationally, therefore, coordinate pairs are treated by the HP 48 exactly like complex numbers. This will be very convenient when working with analytic geometry, as you'll see in Chapter 6.

The polar representation of complex numbers is computed by using the same parametric transformation as with coordinates: $\begin{array}{l} x = r\cos\theta \\ y = r\sin\theta \end{array}$. Algebraically, then, a polar complex number, z, looks like this:

$$z = r(\cos\theta + i\sin\theta) \qquad \text{or} \qquad z = r\,\text{cis}\theta \text{ , for short}$$

Plotting Polar Functions

One of the fifteen kinds of plots built into the HP 48 is designed for plotting functions in polar coordinates. To use it, you need the following information:

- A function, f(θ), expressed in polar form

- The range of θ that you wish to plot

- The horizontal and vertical ranges of the area of the plot you want to view.

- The interval angle between two plotted points—the *resolution* of the plot (higher resolution requires more plotting time).

Example: Plot the polar function, $f(\theta) = \dfrac{6\sin\theta\cos\theta}{\sin^3\theta + \cos^3\theta}$, and find its period.

1. Begin the plot application, ⮕PLOT, and select **Polar** as the plot type.

2. Reset the plot: press DEL ▼ ENTER.

3. Highlight the **EQ:** field, press ⬅EQUATION and type in the right-hand side of the equation above. Press ENTER when finished to insert the function into the **EQ:** field.

4. Change the **INDEP:** variable to **θ**.

5. Notice the angle mode (in the ∡ field). If it's set to **Deg**, then you will want to probably want to use the plotting range 0 to 360, (unless you have a clear idea of the period of the function). If it's set to **Rad**, then use the plotting range 0 to 6.2832 (approximately 2π). Press **OPTS** (or NXT **OPTS**, if necessary) and enter the plotting range for the independent variable given the current angle mode.

6. How often should a point be plotted (the **STEP** value)? The default setting is one point every two degrees (π/90 radians). Twice this resolution (one point every degree or every π/180 radians) often gives a very pleasing plot—although it takes a bit longer to plot. Use the default for now.

7. Press ██ OK ██ ERASE DRAW to plot the function:

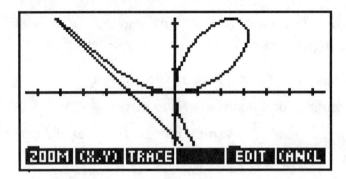

Note that the negatively sloped line on the left-side of the plot appeared almost instantaneously, whereas the other points were plotted individually. Whenever you see this, it indicates the probable presence of an *asymptote*. The curve is called the *Folium of Descartes*. It has the curious property that the area enclosed in the loop is exactly equal to the *non-enclosed* area between the curve and its asymptote!

8. Confirm the presence of an asymptote by replotting the curve with the **CONNECT** feature *off* (unchecked): Press (CANCEL) OPTS (▼) ✓CHK OK ERASE DRAW.

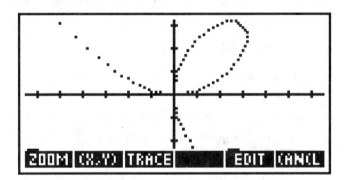

9. Now find the period: Press **TRACE** **(X,Y)** to prepare the cursor to trace along the function while the display indicates the values of θ and the function (Y). Press ▶ to move the cursor "forward" along the curve. The ◀ key moves the cursor "backward" along the curve. As you can see by playing a bit with the cursor, "forward" and "backward" are interpreted as "increasing" and "decreasing" the value of the independent variable (θ). To find the period, press the cursor forward until it begins to retrace the curve. At this point, you can see that the period is 180°, or π radians.

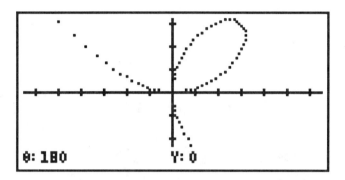

10. Try one more thing: Observe what happens if you move "backwards" along the curve so that θ shows a negative value.... It displays the value of the function even for points outside of the plotting range—and, you will notice, outside of the display range.

Example: Plot the polar equation, $2\cos 3\theta = 2\cos^2\left(\dfrac{\theta}{2}\right)$.

Whenever you plot an equation that includes the independent variable on both sides, you'll get a plot of the left-side expression superimposed upon the plot of the right-side expression. The point(s) of intersection of the two plots represents solution(s) to the equation.

1. Return to the **PLOT** screen: (CANCEL) if you're viewing the display of the plot or (→)(PLOT) if viewing the stack.

2. Highlight the **EQ:** field and enter the equation: (←)(EQUATION)(2)(COS)(3)(α)(→)(F)(▶)(←)(=)(2)(COS)(α)(→)(F)(÷)(2)(▶)(▶)(yˣ)(2)(ENTER).

3. Make sure that the **INDEP** variable is **θ** and that the **H-VIEW** and **V-VIEW** ranges are the defaults.

4. Press **OPTS** and turn on **CONNECT** mode and **SIMULT**aneous plotting mode (not required, just more interesting).

5. Press **OK** **ERASE** **DRAW** to draw the plot of the equation (or, rather, the plots of the two expressions).

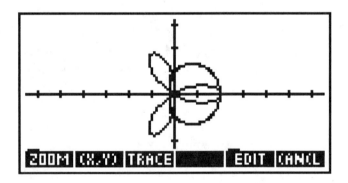

6. Press **TRACE** **(X,Y)** and explore the two expressions. To make the cursor "jump" between the two plots, use the (▲) and (▼) keys. Notice that the points of intersection occur at $\theta = 0° + 360°n$, but that the period of the left-hand side is 180° while that of the right-hand side is 360°.

Example: Find all the points where the two cardioids, $r = 2(1 - \cos\theta)$ and $r = 2(1 + \cos\theta)$, intersect. The HP 48's function analysis menu (**FCN**)—with its handy **ISECT**, **ROOT**, **EXTR**, **AREA** commands—is available only for plots of rectangular functions. So, to find the intersection of two polar functions, you must observe the functions as they are plotted together, then manually explore the regions of intersection.

1. Return to the **PLOT** screen and highlight the **EQ:** field.

2. Enter a list containing the two expressions: ⟵ **{} '** 2 × ⟵ **()** 1 — **COS** ϑ ▶▶▶ **'** 2 × ⟵ **()** 1 + **COS** ϑ **ENTER**.

3. Leave the rest of the plot parameters as they were for the previous example; press **ERASE DRAW**. *Watch the plot* as it's drawn.

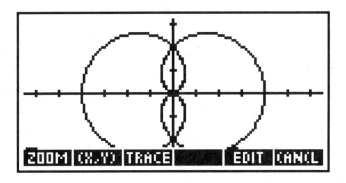

Although there are three points where the *plots* of the two expressions intersect, only two represent true intersections ("collisions") where one value for the independent variable (ϑ) gives the identical value for both expressions. Thus, although each plot contains the origin (0,0), they don't "collide" there.

4. Use the trace feature to find the points of intersection. Press **TRACE** **(X,Y)**, then move forward along one of the cardioids (press and hold ▶) until the cursor is on the upper point of intersection. The display indicates that ϑ: 90 and Y:2. To confirm this as a point of intersection, press ▲ to jump to the other cardioid.... Voilá! The identical coordinates are displayed. Repeating this procedure shows that ϑ:270 and Y:-2 are the coordinates of the other point of intersection.

Plotting Parametric Functions

Two-dimensional functions described parametrically have their own plot type on the HP 48. To use it, you need the following information:

- A function that is described parametrically—i.e. described as a *set* of functions, $x(t)$ and $y(t)$, where the horizontal and vertical coordinates are expressed separately as a function of some parameter t.

- The range of the parameter, t, that you wish to plot.

- The horizontal and vertical ranges within which you want to view the plot.

- The interval step between two successive plotted values of the parameter t. This determines the resolution of the plot.

The Parametric plot type can be confusing at first, because of its relationship to complex numbers. On the HP 48, functions of real numbers are plotted using the Function plot type, but functions of complex numbers are plotted using the Parametric plot type (see also pages 72-75), because a complex number is composed of two parts—like the two coordinates in a parametric representation. But this association of parametric functions with complex numbers means that *you must enter parametric functions as complex numbers*—either in coordinate form , '(x(t), y(t))', or algebraic form, 'x(t)+y(t)*i'.*

Example: Plot the Folium of Descartes using its parametric description:

$$x(t) = \frac{6t}{1+t^3} \quad y(t) = \frac{6t^2}{1+t^3}$$

1. Open the **PLOT** application and change the plot type to Parametric.

2. With the **EQ:** field highlighted, enter the parametric functions together as a single complex number: ⏎EQUATION⏎( )▲6α⏎ T▶1+α⏎Ty×3▶▶⏎,▲6α⏎Ty×2▶▶1 +α⏎Ty×3.

*Note that no matter which form you choose to enter it, the complex-valued function will be displayed in the form determined by the current state of flag -27. If flag -27 is clear (default), it will be displayed in coordinate form. If flag -27 is set, it will be displayed in algebraic form.

$$\left(\frac{6 \cdot t}{1+t^3}, \frac{6 \cdot t^2}{1+t^{30}} \right)$$

Press (ENTER) to return the expression to the **EQ:** field.

3. Now highlight the **INDEP:** field and enter the name of the parameter (**t**), which is the independent variable.

4. Open the **PLOT OPTIONS** screen. The most difficult aspect of plotting a parametric function is pre-determining the appropriate range and step-size of the parameter, but fortunately, the graphing technology of the HP 48 makes it easy to refine your choices. To begin, just use the default step-size (you may need to reset it: highlight the **STEP:** field and press (DEL)(ENTER)). Since the parameter, **t**, is often regarded as "time" when working with real-world applications, try a plotting range of **LO: 0** to **HI: 10**, as in 0 to 10 seconds.

5. Press **OK ERASE DRAW** to draw the plot.

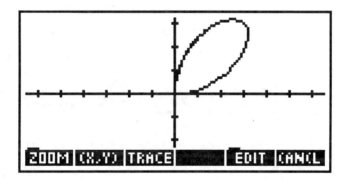

6. You know from the previous section that the graph of the Folium of Descartes includes two asymptotic wings in addition to the loop captured in the plot above. Why didn't they show up in the parametric plot? Explore the plot using the TRACE feature to see if you can find why the wings are "hiding." Press **TRACE** **(X,Y)** and then ▶ repeatedly to move the cursor "forward" along the curve. Notice that, while the cursor moves rapidly at first, it slows to a crawl on the left-side of the loop. Move the cursor beyond **T:10**—points need not be plotted for their coordinates to be displayed. It appears that as t gets larger, the curve approaches the origin asymptotically—but it never sprouts the missing wings. Positive t serves only to define the loop. What about negative t ?

7. Press ◀ and move the cursor so that it first retraces the loop backwards, then moves into a region where t is negative…. Aha! Like a ghost, there are the hidden wings of the Folium of Descartes.

8. Now that you know some key information about the plot, return to the **PLOT OPTIONS** screen ((CANCEL) **OPTS**) and enter a better plotting range, say -10 to 10, so that all of the key features of the plot are drawn. Press **OK** **ERASE** **DRAW** when ready to redraw.

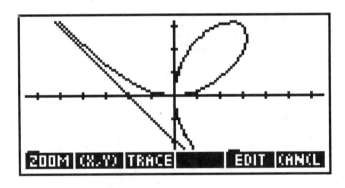

Note how important the kind of parameter is when you try to set a good plotting range: The parameter used in the previous example was *linear*; it needed the entire set of real numbers, $-\infty$ to $+\infty$, to fully describe the graph just *once*. By contrast, the Polar plot-type parameter is an *angle*, which repeats itself as it travels around. Thus, an entire plot is described in just one cycle; extra cycles simply repeat it. But a Parametric plot type also can use an angle as its parameter….

Example: Plot the function defined by:

$$x = \frac{9\cos\theta - 3\cos 9\theta}{5} \qquad\qquad y = \frac{9\sin\theta - 3\sin 9\theta}{5}$$

1. Return to the **PLOT** screen and highlight the **EQ:** field.

2. Enter the parametric functions as a complex number: ⊢EQUATION
   ⊢()▲9COSα→F▶–3COS9α→F▶▶5▶⊢,
   ▲9SINα→F▶–3SIN9α→F▶▶5ENTER.

3. Change the independent variable to the parameter, **θ**.

4. Change the **H-VIEW** range to −18 to 18 and the **V-VIEW** range to
   −10 to 10. (Lucky guesses? Nope—trial and error.)

5. Because the parameter is an angle, you must change the plotting
   range and step interval. Press **OPTS** and change the plotting range
   to 0 to 360 (if in Deg mode) or 0 to 6.2832 (if in Rad mode)
   and change the step interval to 1 (if in Deg mode) or .0174533
   (if in Rad mode).

6. Press **OK** **ERASE** **DRAW** to draw the plot:

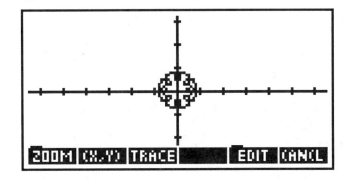

Sure enough, one cycle through the possible angles is sufficient to
draw the complete curve. This particular curve is called a *prolate
epicycloid* and is part of a family of curves generated by a point a
given distance from the center of a small circle rolling around the
outside of a larger circle. See the section on page 76 for more information about these and other interesting curves.

Another advantage to parametric representation is that the vertical and horizontal components of the function can be more easily analyzed separately.

Example: Consider motion that is constrained to a straight line even though the forces controlling the movement are *not* linear, such as the conversion of a circular flywheel motion into the linear motion of a piston:

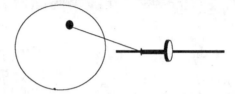

Suppose the motion of a particle moving only along the line $y = 2$, is subjected to nonlinear forces such that the x-coordinate motion is: $x(t) = 2t^3 - 14t^2 + 22t - 9$, where t is time in seconds.

1. Return to the **PLOT** screen and highlight the **EQ:** field.

2. Enter the parametric function as a complex number: $\boxed{\leftarrow}$$\boxed{\text{EQUATION}}$ $\boxed{\leftarrow}$$\boxed{()}$$\boxed{2}$$\boxed{\alpha}$$\boxed{\leftarrow}$$\boxed{T}$$\boxed{y^x}$$\boxed{3}$$\boxed{\triangleright}$$\boxed{-}$$\boxed{1}$$\boxed{4}$$\boxed{\alpha}$$\boxed{\leftarrow}$$\boxed{T}$$\boxed{y^x}$$\boxed{2}$$\boxed{\triangleright}$$\boxed{+}$$\boxed{2}$$\boxed{2}$$\boxed{\alpha}$$\boxed{\leftarrow}$$\boxed{T}$ $\boxed{-}$$\boxed{9}$$\boxed{\leftarrow}$$\boxed{,}$$\boxed{2}$$\boxed{\text{ENTER}}$. Notice that the y-component is a constant, 2.

3. Change the **INDEP:** variable to **t**.

4. Set the **H-VIEW** to −20 to 50 and the **V-VIEW** to −2 to 10.6.

5. Set the plot range (in **PLOT OPTIONS**) to 0 to 6 and **STEP:** to .05.

6. Draw the plot (**OK** **ERASE** **DRAW**) and *watch as it plots*:

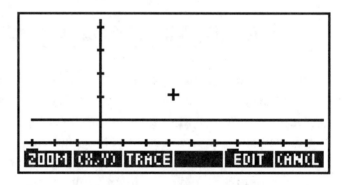

The plot is a straight-line, of course, because the y-component is constrained to be a constant. But did you notice how it was created?

7. Use the Trace feature to explore the function. Press $\boxed{\text{TRACE}}$ $\boxed{\text{(X,Y)}}$ and then $\boxed{\blacktriangleright}$ repeatedly to increase the value of the parameter in steps of .05 seconds. The cursor moves to the right then seems to pause and return back to the left, then pauses again and moves back to the right—in apparent retrograde motion.

Example: Add a second parametric function to that in the previous example — identical except for the *y*-component, which should be $y(t) = t$.

1. Return to the **PLOT** screen and highlight the **EQ:** field.

2. You want to add a second parametric function to the one already in place. Use the Calc feature to get access to the stack, where you can copy the current function, edit the copy, and combine the original and modified versions together into a list: $\boxed{\text{NXT}}$ $\boxed{\text{CALC}}$ $\boxed{\text{ENTER}}$ $\boxed{\Leftarrow}$ $\boxed{\text{EDIT}}$ $\boxed{\blacktriangledown}$ $\boxed{\blacktriangleright}$ $\boxed{\blacktriangleright}$ $\boxed{\blacktriangleright}$ $\boxed{\text{DEL}}$ $\boxed{\alpha}$ $\boxed{\Leftarrow}$ $\boxed{\text{T}}$ $\boxed{\text{ENTER}}$ $\boxed{2}$ $\boxed{\text{PRG}}$ $\boxed{\text{LIST}}$ $\boxed{\text{→LIST}}$.

3. Enter the list into the **EQ:** field: $\boxed{\Leftarrow}$ $\boxed{\text{CONT}}$ $\boxed{\text{OK}}$.

4. Redraw the plot: $\boxed{\text{NXT}}$ $\boxed{\text{ERASE}}$ $\boxed{\text{DRAW}}$:

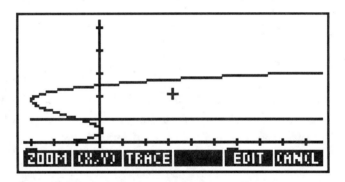

Now you can see the "hidden" function controlling the movement of the *x*-component because the function has been spread out by allowing *y* to vary with time. Press $\boxed{\text{TRACE}}$ and use the arrow keys to explore the relationship visually.

Plotting Functions of Complex Numbers

As mentioned earlier, the Parametric plot type on the HP 48 is actually a general purpose complex function plot type. This is why parametric functions are entered as complex numbers—or complex-valued functions, to be precise.

A complex-valued function, $f(x+iy) = (u+iv)$ takes a complex number (x,y) and maps it to the complex number (u,v). In order to plot the function of the complex number, you must first determine u and v.

Example: For the function $f(z) = z^2$, where z is the complex number $(x+iy)$, compute the complex number (u,v).

1. Make sure flag -27 is clear (press [2][7][+/−][α][C][α][F][ENTER]) and enter the symbolic complex number $'(\times, y)'$ onto the stack.

2. Square it: [2][y^x].

3. Symbolically expand and collect the result: [←][SYMBOLIC] **EXPA** **EXPA** **COLCT**. Result: $'(-y^2+x^2, 2*x*y)'$

Therefore, your complex-valued function is $\quad u = x^2 - y^2 \quad v = 2xy$

Notice that you *cannot* simply plot this result—the complex-valued function $'x^2-y^2+2*x*y*i'$—using the Parametric plot type, because you have two "independent" variables (x and y) instead of one (usually t or θ). However, you may plot this function if either x or y is given a value....

Example: Plot $f(z) = z^2$, where z is the complex number $(x+3i)$.

1. Open the **PLOT** screen and make sure that the plot type is still set to **Parametric**. Highlight the **EQ:** field.

2. Press (NXT) **CALC** to move to the stack. Now enter the complex number onto the stack: (')(←)(())(α)(←)(X)(←)(,)(3)(ENTER).

3. Square the complex number, then expand and collect the result as you did in the previous example: (2)(y^x)(←)(SYMBOLIC) **EXPA** **EXPA** **COLCT**.

    Result: $'(x^2-9, 6*x)'$.

4. Return the result to the **EQ:** field: (←)(CONT) **OK** (NXT).

5. Change the **INDEP:** variable to **X**.

6. Switch to the **PLOT OPTIONS** screen (**OPTS**) and change the plotting range to -10 to 10 and the **STEP** to $.5$. Press **OK** .

7. Change **H-VIEW** to -13 to 50 and check **AUTOSCALE**.

8. Draw the plot: **ERASE** **DRAW** .

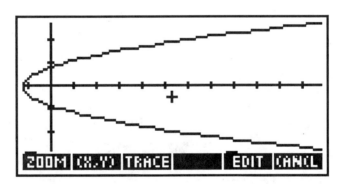

The Parametric plot type allows you only to partially plot a complex function, but the HP 48 also has a plot type, Gridmap, capable of plotting the complete mapping. Basically, a gridmap plots a *series* of parametric curves, allowing y, then x, to vary through a series of steps, as if superimposing a series of parametric plots where, say, $y = 3$, then $y = 2$, then $y = 1$, etc; then $x = 3$, $x = 2$, $x = 1$, etc.

Example: Plot a grid representation of the complex plane, where $f(z) = z$. This mapping is analogous to the real number function $f(x) = x$ (a straight line), except that it results in a rectilinear *plane* instead of line. The Gridmap plot type represents this plane as a grid—plotting only a few of the infinite number of lines of the plane and allowing those lines to stand for the plane as a whole.

1. At the **PLOT** screen, highlight the **TYPE:** field and change it to Gridmap: α G (Gridmap is the only plot type starting with G).

2. Move the highlight to the **EQ:** field and enter the symbolic complex number ' (x,y) ' . (If flag -27 is set, you will see the complex number displayed as 'x+y*i ').

3. Make sure that the **INDEP:** contains X and **DEPND:** contains y (note the lower-case). In this case, X is the independent variable not because it is any more "independent" than y, but because it is to be plotted on the horizontal axis.

4. Use 10 STEPS for X and 8 STEPS for y. This means that 10 values for X and 8 values for y will be used, so 80 points will be plotted.

5. Draw the plot: ERASE DRAW ⊟.

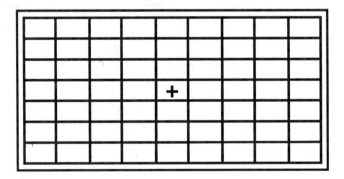

Example: Plot the complex-valued mapping, $f(z) = z^2$, using the Gridmap plot type. What were previously straight lines within the complex plane are now transformed by the function into curves—again, analogous to what happens to a straight-line in the real number plane when it is transformed by a function.

1. Return to the **PLOT** screen, and highlight the **EQ:** field.

2. Modify the expression so that it is ' ⟨×,ㄣ⟩^2': **EDIT** ⇨▷ ◁ ⱻ 2 **OK**.

3. Draw the plot: **ERASE DRAW** ⊖.

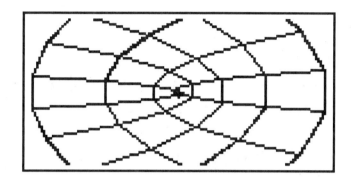

Example: Repeat the previous example using $f(z) = z^3$ as the transformation.

Result:

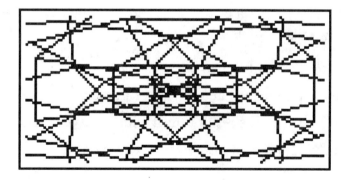

A Garden of Curves

Polar and parametric plotting allow you to view some very interesting curves. The examples in this chapter just barely hint at some of the exploratory (and artistic) possibilities these curves offer.

This section gives you additional fodder for your curiosity. Each entry includes some information about the curve or curve family and an example plotted on the HP 48, along with the plotting parameters used to create the example.

Note: If a parameter isn't referred to, then the default settings are assumed. Also, "Cartesian" refers to a curve's description in a rectangular coordinate system.

Cassinian Curves

A Cassinian curve is the locus of points, P, the product of whose distances from two fixed points, F_1 and F_2, is constant. That is, $PF_1 \cdot PF_2 = k$.

Ovals of Cassini. Here are the forms of the function:

- Cartesian: $\left(x^2 + y^2\right)^2 - 2e^2\left(x^2 - y^2\right) = a^4 - e^4$

- Polar: $r = \sqrt{d^2 \cos 2\theta \pm \sqrt{d^4 \cos^2 2\theta + \left(a^4 - d^4\right)}}$

There are four different shapes for Cassinian ovals. The shape depends on relationship of a (the square root of the constant k) and e (half of the distance between the two fixed points):

When $a < e$, the result is two oval islands.

When $a = e$, the result is Lemniscate of Bernoulli (see below).

When $e < a < e\sqrt{2}$, the result is oval with concave sides.

When $a > e\sqrt{2}$, the result is oval with convex sides.

Example: EQ: {'√(d^2*COS(2*θ)+√(d^4*COS(2*θ)^2
+a^4-d^4))''√(d^2*COS(2*θ)-
√(d^4*COS(2*θ)^2+a^4-d^4))'}

TYPE: Polar ∠: Deg INDEP: θ
H-VIEW: -2 2 V-VIEW: -1.1 1.1
LO: 0   HI: 360 STEP: 1

$a = .9; d = 1$

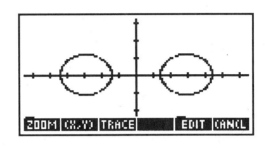

Example: $a = 1.1; d = 1$

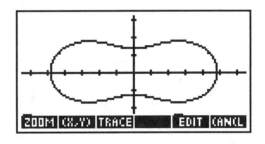

Example: $a = 1.5; d = 1$

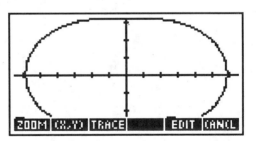

Lemniscate of Bernoulli. The *Lemniscate of Bernoulli* is a special case of a Cassinian oval, where $a = d$. The area of one of the loops is a^2.

- Cartesian: $\left(x^2 + y^2\right)^2 = 2a^2\left(x^2 - y^2\right)$

- Polar: $r^2 = 2a^2 \cos 2\theta$

- Parametric:
$$x = \frac{at\sqrt{2}\left(1 + t^2\right)}{1 + t^4}$$

$$y = \frac{at\sqrt{2}\left(1 - t^2\right)}{1 + t^4}$$

Example: EQ: `'√(2*a^2*COS(2*θ))'`
TYPE: Polar   ∡: Deg   INDEP: θ
H-VIEW: −3 3   V-VIEW: −1.5 1.5
LO: 0   HI: 360   STEP: 1

$a = 1.5$

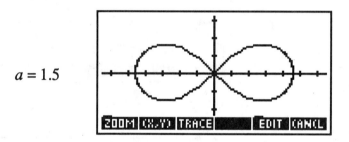

       3. POLAR AND PARAMETRIC EQUATIONS

Cissoids and Conchoids

A *cissoid* is the locus of points P that are the same distance from a fixed point, F_1, as the distance between points, Q and R, on two curves such that F_1, P, Q, and R are collinear. Ordinary cissoids employ a circle and a line as the two curves.

A *conchoid* is the locus of points, P_1 and P_2, that are equidistant from a point Q on a given curve and a fixed point F_1 along a line containing both Q and F_1. If you draw a line from F_1 through Q, then P_1 will be a distance k farther from F_1 than Q and P_2 will be a distance k closer to F_1 than Q along the drawn line. Conchoids are cousins of cissoids, a fact which becomes clearer when you consider that the *Conchoid of Nicomedes* (discussed below) is both an ordinary cissoid and the conchoid of a line with respect to a fixed point not on the line.

Cissoid of Diocles. The *Cissoid of Diocles* is an ordinary cissoid with the origin as the fixed point, the point $(R,0)$ as the center of the circle (radius $= R$), and the line $x = 2R$ as the line. The curve is asymptotic to the line $x = 2R$; and the area between curve and asymptote is $3R^2\pi$.

- Cartesian: $y^2 = \dfrac{x^3}{2R - x}$

- Parametric: $x = \dfrac{2Rt^2}{1+t^2}$ $y = \dfrac{2Rt^3}{1+t^2}$

- Polar: $r = 2R\sin\theta\tan\theta$

Example: EQ: '2*R*SIN(θ)*TAN(θ)'

TYPE: Polar   ∡: Deg   INDEP: θ
H-VIEW: −6.5 6.5   Y-VIEW: −3.1 3.2
LO: 0   HI: 360   STEP: 1

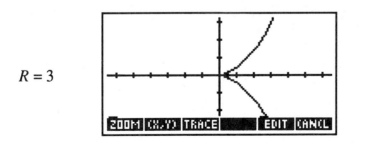

$R = 3$

Folium of Descartes. The *Folium of Descartes* is a cissoid of the ellipse defined by $x^2 - xy + y^2 - a(x+y) = 0$, and the straight line $y = -x - a$. The curve is asymptotic to the line $y = -x - a$; the vertex of loop is at $(3a/2, 3a/2)$; the area of loop is $\dfrac{3a^2}{2}$, as is also the area between curve and its asymptote.

- Cartesian: $x^3 + y^3 - 3axy = 0$

- Polar: $r = \dfrac{3a \sin\theta \cos\theta}{\sin^3\theta + \cos^3\theta}$

- Parametric: $x = \dfrac{3at}{1+t^3} \qquad y = \dfrac{3at^2}{1+t^3}$

Example: EQ: '(3*a*SIN(θ)*COS(θ))/(SIN(θ)^3
          +COS(θ)^3)'

TYPE: Polar   ∡: Deg   INDEP: θ
H-VIEW: −14 14   V-VIEW: −7 7
LO: 0   HI: 180   STEP: 2

$a = 4$

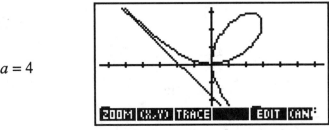

Conchoid of Nicomedes. The *Conchoid of Nicomedes* is an ordinary cissoid with the fixed point being the center of the circle. The curve has an asymptote at $x = a$, which lies between the y-axis and cissoid's second "curve," the line $x = a + b$. The shape depends upon the relation of a and b.

- Cartesian: $$\left(x^2 + y^2\right)(x - a)^2 - b^2 x^2 = 0$$

- Polar: $$r = \frac{a}{\cos\theta} \pm b$$

- Parametric: $$\begin{aligned} x &= a + b\cos\theta \\ y &= a\tan\theta + b\sin\theta \end{aligned}$$

Example: EQ: `'(a+b*COS(θ),a*TAN(θ)+b*SIN(θ))'`
TYPE: Parametric   ∡: Deg   INDEP: θ
H-VIEW: -6.5 6.5   V-VIEW: -3.1 3.2
LO: 0   HI: 360   STEP: 1

$a = 3; b = 2$

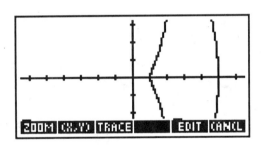

Example: $a = 2; b = 3$

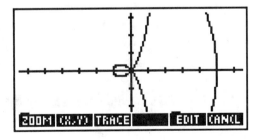

Strofoid. A *strofoid* is a cissoid of a circle (radius = a) and a line through its center with respect to a fixed point on the circle. The vertex of the loop is at $(a,0)$; the loop area is $a^2\left(1-\dfrac{\pi}{4}\right)$; the area between curve and asymptote is $a^2\left(1+\dfrac{\pi}{4}\right)$.

- Cartesian: $y^2 = x^2\dfrac{a-x}{a+x}$

- Polar: $r = a\dfrac{\cos 2\theta}{\cos\theta}$

- Parametric: $x = \dfrac{a(1-t^2)}{1+t^2}$ $y = \dfrac{at(1-t^2)}{1+t^2}$

Example: EQ: 'a*COS(2*θ)/COS(θ)'

TYPE: Polar   **∡: Deg   INDEP: θ**
H-VIEW: -6.5 6.5   **V-VIEW: -3.1 3.2**
LO: 0   **HI: 360   STEP: 1**

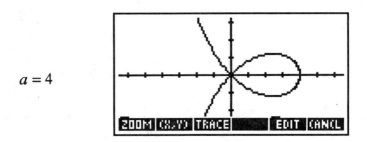

$a = 4$

Pascal's Snails. Pascal's snails, or *limaçons*, are conchoids of a circle where the fixed point is on the circle—i.e. the locus of points P_1 and P_2 a distance b from each point on a circle of diameter a, as measured along a line containing a fixed point on the circle. Limaçons come in four typical shapes, depending on the relation of a to b. When $b \geq a$, the area enclosed by the curve is $\pi\left(b^2 + \dfrac{a^2}{2}\right)$

- Cartesian: $x^2 + y^2 - ax^2 - b^2(x^2 + y^2) = 0$

- Polar: $r = a\cos\theta \pm b$

- Parametric: $\begin{aligned} &x = a\cos^2\theta + b\cos\theta \\ &y = a\cos\theta\sin\theta + b\sin\theta \end{aligned}$

3. Polar and Parametric Equations

Example: EQ: '(a*COS(θ)^2+b*COS(θ),
a*COS(θ)*SIN(θ)+b*SIN(θ))'

TYPE: Parametric   ∠: Rad   INDEP: θ
H-VIEW: −15 15   V-VIEW: −7.2 7.4
LO: 0   HI: 6.3   STEP: .05

$a = 4, b = 5$ $(a < b < 2a)$

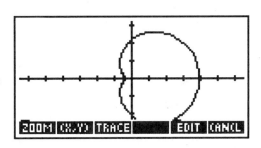

Example: $a = 2, b = 5$ $(b > 2a)$

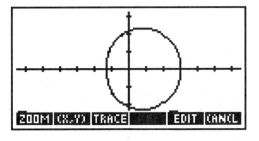

Example: $a = b = 5$ $(a = b)$

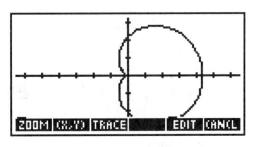

Example: $a = 5, b = 2$ $(a > b)$

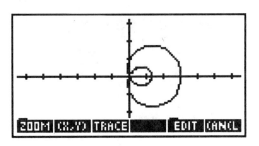

A Garden of Curves

Cycloidal Curves

Cycloidal curves are an interesting family of curves. They represent the motion of a point on, "beyond" or "within" a circle as it rolls along another curve. The *cycloid* itself represents the motion of a point on a circle as it rolls along a straight line. The *epicycloid* represents the motion of a point on a circle as it rolls along the outside of a second circle. The *hypocycloid* represents the motion of a point on a circle as it rolls along the inside of a second circle. Then, for each of those three branches of the family, there are the *trochoid* versions, where the point on the circle isn't precisely *on* the circle but "beyond" the radius of the circle (*prolate*) or "within" the radius of the circle (*curtate*).

The learning toy, *Spirograph*™, makes extensive use of the cycloidal family of curves, with circles of differing radii rolled around or within one another to form beautiful patterns. In particular, *Spirograph* utilizes the key feature of the cycloid curves—the *ratio* of the radii (as expressed by the number of "teeth") on two circles. The program, SPIRO, on page 312, takes the number of teeth of the fixed circle from level 3, the number of teeth of the rolling circle from level 2, and a number indicating whether it rolls inside (-1) or outside (1) the fixed circle. SPIRO always draws a prolate curve.*

Ordinary Cycloid. The ordinary cycloid is generated by a fixed point P on a circle of radius a which rolls without slipping along the x-axis. The period of curve is $2\pi a$; the length of curve between two cusps is $8a$; the area between one full arch of the curve and x-axis is $3\pi a^2$.

- Cartesian:
$$x = a\cos^{-1}\frac{a-y}{a} - \sqrt{y(2a-y)}$$

- Parametric:
$$x = a(\theta - \sin\theta)$$
$$y = a(1 - \cos\theta)$$

Example: EQ: '(a*(θ-SIN(θ)),a*(1-COS(θ)))'
TYPE: Parametric   ∡: Rad   INDEP: θ
H-VIEW: -30 30   Y-VIEW: -2.5 15
LO: -10   HI: 10   STEP: .05

*Short programs like SPIRO may be written to provide easy control of parameters for other curve families as well).

$a = 3$

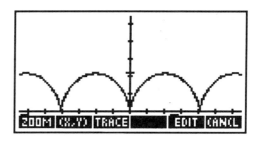

Trochoid. The trochoid curve is generated by a fixed point P at a distance λa from the center of a circle of radius a which rolls without slipping along the x-axis. If $\lambda < 1$, curve is *curtate cycloid*. The "base" is the horizontal line above x-axis. If $\lambda > 1$, curve is *prolate cycloid*. The "base" is the horizontal line below x-axis. If $\lambda = 1$, curve is ordinary cycloid.

- Parametric: $x = a(\theta - \lambda \sin \theta)$ $\qquad$ $y = a(1 - \lambda \cos \theta)$

Example: EQ: '(a*(θ-λ*SIN(θ)),a*(1-
λ*COS(θ)))'

TYPE: Parametric $\quad$ ∡: Rad $\quad$ INDEP: θ
H-VIEW: -30 30 $\qquad$ Y-VIEW: -5 12
LO: -10 $\qquad\qquad$ HI: 10 $\quad$ STEP: .05

$a = 3;\ \lambda = 0.5$

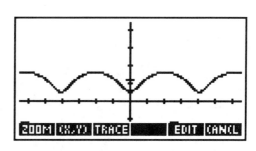

Example: $a = 3;\ \lambda = 1.5$

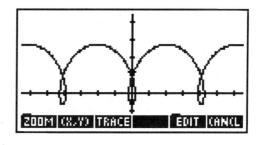

Ordinary Epicycloid. The ordinary epicycloid is generated by a fixed point P on a circle of radius b, which rolls without slipping on the outside of a fixed circle of radius a. If $a/b=N$ is an integer, then: the curve has N equal branches; the arc length of each branch is $\dfrac{8}{N}(a+b)$; the area of one sector is $\dfrac{b\pi}{a}(a+b)(a+2b)$.

- Parametric:
$$x = (a+b)\cos\theta - b\cos\frac{a+b}{b}\theta$$
$$y = (a+b)\sin\theta - b\sin\frac{a+b}{b}\theta$$

Example:
```
EQ: '((a+b)*COS(θ)-b*COS((a+b)/b*θ),
    (a+b)*SIN(θ)-b*SIN((a+b)/b*θ))'
TYPE: Parametric      ∡: Rad         INDEP: θ
H-VIEW: -15 15        V-VIEW: -7.2 7.4
LO: 0                 HI: 6.3        STEP: .05
```

$a = 5; b = 1$
(integral ratio)

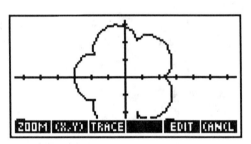

Example:
```
EQ: '((a+b)*COS(θ)-b*COS((a+b)/b*θ),
    (a+b)*SIN(θ)-b*SIN((a+b)/b*θ))'
TYPE: Parametric      ∡: Rad         INDEP: θ
H-VIEW: -24 24        V-VIEW: -12 12
LO: 0                 HI: 19         STEP: .05
```
Reduce the ratio a/b, if possible; and plot one cycle (2π) for each b.

$a = 5; b = 3$
(non-integral ratio)

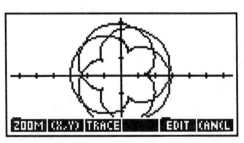

Nephroid. The nephroid is the 2-cusped epicycloid ($a = 2b$).

Example: EQ: '((a+b)*COS(θ)-b*COS((a+b)/b*θ),
          (a+b)*SIN(θ)-b*SIN((a+b)/b*θ))'

TYPE: Parametric   ∡: Rad   INDEP: θ
H-VIEW: -15 15   V-VIEW: -7.2 7.4
LO: 0   HI: 6.3   STEP: .05

$a = 2; b = 1$

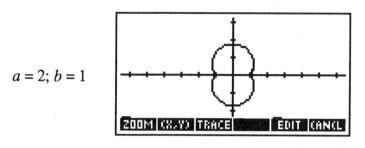

Cardioid. The *cardioid* is an ordinary epicycloid where two circles are the same size ($a = b$), which simplifies the epicycloid equation. The length of the curve is $16a$; the area enclosed by the curve is $7a^2\pi$.

- Cartesian: $\left(x^2 + y^2 - a^2\right)^2 = 4a^2\left[(x-a)^2 + y^2\right]$

- Polar: $r = 2a(1 - \cos\theta)$

- Parametric: $x = a(2\cos\theta - \cos 2\theta)$
                 $y = a(2\sin\theta - \sin 2\theta)$

Example: EQ: '2*a*(1-COS(θ))'

TYPE: Polar   ∡: Rad   INDEP: θ
H-VIEW: -30 30   V-VIEW: -15 15
LO: 0   HI: 6.3   STEP: .05

$a = 5$

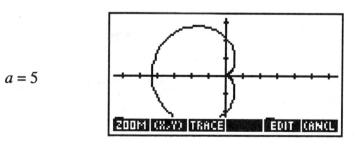

Epitrochoid. The epitrochoid is generated by a fixed point P at a distance $b\lambda$ from the center of a circle of radius b, which rolls without slipping on the outside of a fixed circle of radius a. If $\lambda < 1$, curve is a *curtate epicycloid*. If $\lambda > 1$, curve is a *prolate epicycloid*. If $\lambda = 1$, curve is a normal cycloid. If $a/b=N$ is an integer, the curve consists of N equal branches. If N is a fraction, the branches intersect.

$$x = (a+b)\cos\theta - b\lambda\cos\frac{a+b}{b}\theta$$

• Parametric:

$$y = (a+b)\sin\theta - b\lambda\sin\frac{a+b}{b}\theta$$

Example: EQ: '((a+b)*COS(θ)-b*λ*COS((a+b)/
            b*θ),(a+b)*SIN(θ)-b*λ*SIN((a+b)/
            b*θ))'

TYPE: Parametric   ∡: Rad   INDEP: θ
H-VIEW: -30 30   V-VIEW: -15 15
LO: 0   HI: 6.3   STEP: .05

$a = 5; b = 1; \lambda = 0.5$

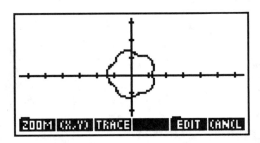

Example: $a = 5; b = 1; \lambda = 2$

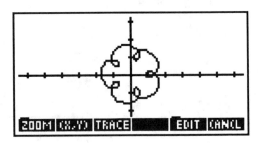

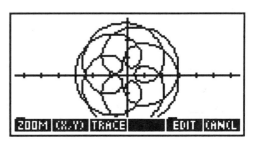

Example: $a = 5; b = 3; \lambda = 2$

Ordinary Hypocycloid. The ordinary *hypocycloid* is generated by a fixed point P on a circle of radius b which rolls without slipping on the inside of a fixed circle of radius a. If $a/b=N$ is an integer, the curve has N equal branches; if N is a fraction, the branches cross one another. If $N = 2$, the hypocycloid reduces to a straight line.

The arc length of each branch is $\dfrac{8}{N}(a - b)$; the area of a sector is $\dfrac{b\pi}{a}(a - b)(a - 2b)$.

- Parametric:

$$x = (a - b)\cos\theta + b\cos\frac{a - b}{b}\theta$$

$$y = (a - b)\sin\theta - b\sin\frac{a - b}{b}\theta$$

Example: `EQ: '((a-b)*COS(θ)+b*COS((a-b)/`
`            b*θ),(a-b)*SIN(θ)-b*SIN((a-b)/`
`            b*θ))'`

`TYPE: Parametric      ∠: Rad        INDEP: θ`
`H-VIEW: -12 12        V-VIEW: -6 6`
`LO: 0                 HI: 6.3        STEP: .05`

$a = 5; b = 1$

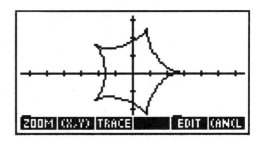

Astroid. The *astroid* is a hypocycloid with 4 cusps ($a = 4b$)—a simpler equation than the general hypocycloid. The length of the curve is $6a$; the area between the curve and the fixed circle is $\dfrac{5}{8}a^2\pi$; the area enclosed by the curve is $\dfrac{3}{8}a^2\pi$.

- Cartesian: $\qquad x^{\frac{2}{3}} + y^{\frac{2}{3}} = a^{\frac{2}{3}}$
- Parametric: $\qquad x = a\cos^3\theta \qquad y = a\sin^3\theta$

Example: EQ: '(a*COS(θ)^3,a*SIN(θ)^3)'

TYPE: Parametric   ∡: Rad   INDEP: θ
H-VIEW: −12 12   Y-VIEW: −6 6
LO: 0   HI: 6.3   STEP: .05

$a = 5$

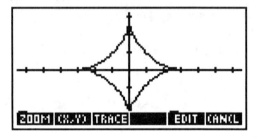

Deltoid. A *deltoid* is a 3-cusped hypocycloid ($a = 3b$)—a simpler equation:

- Parametric: $\qquad x = b(2\cos\theta + \cos2\theta) \quad y = b(2\sin\theta - \sin2\theta)$

Example: EQ: '(b*(2*COS(θ)+COS(2*θ)),
        b*(2*SIN(θ)−SIN(2*θ)))'

TYPE: Parametric   ∡: Rad   INDEP: θ
H-VIEW: −12 12   Y-VIEW: −6 6
LO: 0   HI: 6.3   STEP: .05

$b = 2$

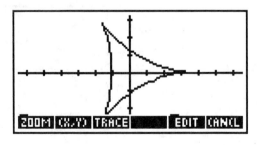

Hypotrochoid. The *hypotrochoid* is generated by a fixed point P at a distance $b\lambda$ from the center of a circle of radius b which rolls without slipping on the inside of a fixed circle of radius a. If $\lambda < 1$, curve is a *curtate hypocycloid*. If $\lambda > 1$, curve is a *prolate hypocycloid*. If $\lambda = 1$, curve is a normal cycloid. If $a/b=N$ is an integer, the curve consists of N equal branches; if N is a fraction, the branches intersect.

- Parametric:

$$x = (a - b)\cos\theta + b\lambda\cos\frac{a - b}{b}\theta$$

$$y = (a - b)\sin\theta - b\lambda\sin\frac{a - b}{b}\theta$$

Example: EQ: `'((a-b)*COS(θ)+b*λ*COS((a-b)/`
`b*θ),(a-b)*SIN(θ)-b*λ*SIN((a-b)/`
`b*θ))'`

TYPE: Parametric   ∡: Rad   INDEP: θ
H-VIEW: -12 12   V-VIEW: -6 6
LO: 0   HI: 6.3   STEP: .05

$a = 5; b = 1; \lambda = 0.5$

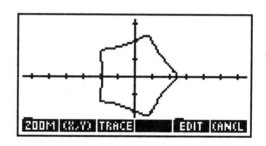

Example: $a = 5; b = 1; \lambda = 2$

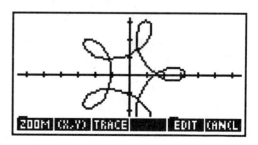

Example: H-VIEW: −90 90   V-VIEW: −45 45
LO: 0   HI: 19

$a = 40; b = 3; \lambda = 3$

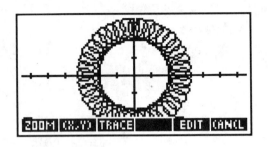

Example: H-VIEW: −90 90   V-VIEW: −45 45
LO: 0   HI: 32

$a = 36; b = 15; \lambda = 0.6$

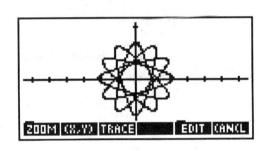

Example: H-VIEW: −106 106   V-VIEW: −53 53
LO: 0   HI: 95

$a = 49; b = 15; \lambda = 1.4$

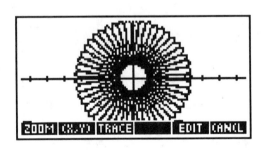

Roses. A *rose* is a hypotrochoid in which $\lambda = \dfrac{a-b}{b}$, which makes for a simple polar form. Note that roses can be either curtate or prolate hypocycloids.

- Polar: $r = 2(a - b)\cos\dfrac{a}{a - 2b}\theta$

Example: EQ: `'2*(a-b)*COS(a/(a-2*b)*θ)'`
TYPE: Polar ∡: Rad INDEP: θ
H-VIEW: −12 12 V-VIEW: −6 6
LO: 0 HI: 6.3 STEP: .05

$a = 5; b = 2$

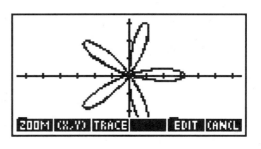

Example: H-VIEW: −2 2 V-VIEW: −1 1
LO: 0 HI: 22

$a = 4; b = 3.5$

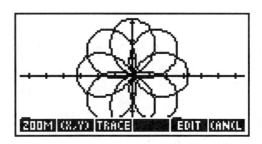

Spirals

Spirals all share certain common traits: the polar radius gets larger as the polar angle increases; and the function is *not* periodic.

Spiral of Archimedes. The polar radius is proportional to the polar angle. The arc length of the curve is $\frac{a}{2}\left(\theta\sqrt{\theta^2+1}+\sinh^{-1}\theta\right)$, which, for large θ, is approximately $\frac{a}{2}\theta^2$; the area of the sector bounded by two radian angles is $\frac{a^2}{6}\left(\theta_2{}^3-\theta_1{}^3\right)$.

- Polar: $r = a\theta$

Example: EQ: `'a*θ'`
TYPE: Polar   ∡: Rad   INDEP: θ
H-VIEW: −150 150   V-VIEW: −75 75
LD: 0   HI: 32   STEP: .05

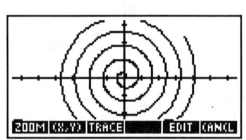

$a = 3$

Hyperbolic Spiral. The *hyperbolic spiral* is the inverse of the Spiral of Archimedes, with an asymptote at $y = a$. It represents the path of a particle under a central force that varies as the cube of the distance of the particle from the central force. The area of sector bounded by two radian angles is $\frac{a^2}{2}\left(\frac{1}{\theta_1}-\frac{1}{\theta_2}\right)$.

- Polar: $r = \dfrac{a}{\theta}$.

Example: EQ: `'a/θ'`
TYPE: Polar   ∡: Rad   INDEP: θ
H-VIEW: −6.5 6.5   V-VIEW: −3.1 3.2
LD: 0   HI: 16   STEP: .05

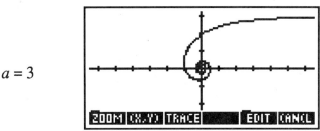

$a = 3$

Logarithmic Spiral. The *logarithmic spiral*, also known as the *equiangular spiral*, is the spiral form often seen in nature—in the Nautilus shell, in the arrangement of sunflower seeds, and in the formation of pine cones. It is "equiangular:" the angle β formed between the tangent to any point P on the spiral and the polar radius (the segment connecting P to the pole) is constant.

Other interesting properties: The length of an arc from the pole along the spiral to r is $\dfrac{r}{\cos\beta}$; and lengths of r drawn at equal angular intervals to each other form a geometric progression; also, if you roll the spiral along a line, the path of the pole is also a line.

- Polar: $r = ae^{\theta/\tan\beta}$

Example: EQ: `'a*e^(θ/TAN(β))'`

TYPE: Polar   ∡: Rad   INDEP: θ
H-VIEW: -35 35   V-VIEW: -17.5 17.5
LO: 0   HI: 20   STEP: .05

$a = 0.1; \beta = 1.3$

Parabolic Spiral. The *parabolic spiral* is so named because of its analogy to the equation for a parabola: $y^2 = a^2 x$

- Polar: $r^2 = a^2 \theta$

Example: EQ: '√(a^2*θ)'

TYPE: Polar   ∡: Rad   INDEP: θ
H-VIEW: −18 18   V-VIEW: −9 9
LO: 0   HI: 40   STEP: .05

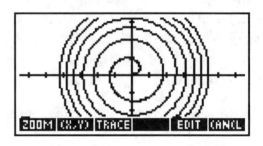

$a = 2$

Lituus Spiral. The *Lituus spiral* is the inverse with respect to the pole of the parabolic spiral. It has the x-axis as an asymptote. It is the spiral often used in the whorls sitting atop columns in classical (Roman) architecture.

- Polar: $r^2 = \dfrac{a^2}{\theta}$

Example: EQ: '√(a^2/θ)'

TYPE: Polar   ∡: Rad   INDEP: θ
H-VIEW: −2.5 6.5   V-VIEW: −2.25 2.25
LO: 0   HI: 20   STEP: .05

$a = 2$

Sinusoidal Spirals. A particle acted upon by a central force that is inversely proportional to the ($2n+3$) power of its distance from the force (where n is a rational number) moves along a sinusoidal spiral. Some special cases:

$n = -2$	rectangular hyperbola		$n = 2$	lemniscate
$n = -1$	line		$n = 1$	circle
$n = -1/2$	parabola		$n = 1/2$	cardioid

- Polar: $\qquad r^n = a^n \cos(n\theta) \qquad$ or $\qquad r^n = a^n \sin(n\theta)$

Example: 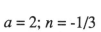 EQ: '(a^n*SIN(n*θ))^(1/n)'
TYPE: Polar   ∡: Rad   INDEP: θ
H-VIEW: −40 40   V-VIEW: −25 15
LO: 0   HI: 9   STEP: .05

$a = 2;\ n = -1/3$

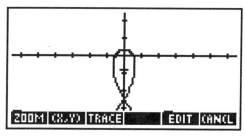

Example: H-VIEW: −5 5   V-VIEW: −2.5 2.5
LO: 0   HI: 30   STEP: .05

$a = 2;\ n = 3/4$

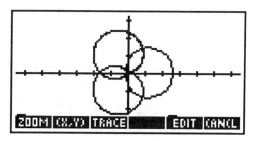

Example: H-VIEW: −20 20   V-VIEW: −10 10
LO: 0   HI: 32   STEP: .05

$a = 2;\ n = -7/5$

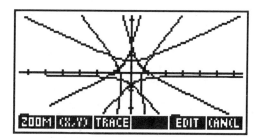

Other Curves

Lissajous. The parametric description for the standard sine curve is: $x=\theta$, $y=a\sin(b\theta+c)$; only the y-component undergoes the sine function. But in the lissajous, both components undergo their own independent sine functions.

- Parametric:
$$x = a\sin(m\theta + c)$$
$$y = b\sin(n\theta + d)$$

Example: EQ: `'(a*SIN(m*θ+c),b*SIN(n*θ+d))'`
TYPE: Parametric   ∡: Rad   INDEP: θ
H-VIEW: −6.5 6.5   V-VIEW: −3.1 3.2
LD: 0   HI: 6.3   STEP: .05

$a = 4$
$b = 3$
$c = 0.5$
$d = 0.8$
$m = 2$
$n = 5$

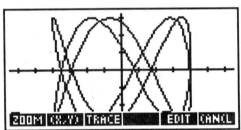

Tractrix. The *tractrix* is the curve of points P such that the distance from P to the x-axis along the tangent at P is constant. It is the track of the back wheel of a bicycle as the front wheel makes a 90° turn.

- Parametric:
$$x = a\ln(\sec\theta + \tan\theta) - a\sin\theta$$
$$y = a\cos\theta$$

Example: EQ: `'(a*LN(1/COS(θ)+TAN(θ))−`
`a*SIN(θ),a*COS(θ))'`
TYPE: Parametric   ∡: Rad   INDEP: θ
H-VIEW: −11.5 11.5 V-VIEW: −1 12
LD: 0   HI: 6.3   STEP: .05

$a = 4$

Witch of Agnesi. The *Witch of Agnesi*, whose name comes from a mistranslation of the Italian *versoria* ("free to move in any direction") as *versiera* ("witch", or "devil's wife"), is a curve that is asymptotic to x-axis, with the area between curve and asymptote equal to $4R^2\pi$. It is an unusual curve, whose definition is somewhat complicated. Here's how to construct the graph manually:

1. From a fixed point F_1 on a circle of radius R, construct the tangents to the circle at F_1 and at F_2, the point on the circle diametrically opposite F_1.

2. Then at an angle θ from the diameter connecting F_1 and F_2, draw a secant from F_1 to point T_2 on the opposite tangent line. The secant intersects the circle at Q.

3. Finally a draw a line through Q that is parallel to the tangents and a line through T_2 parallel to the diameter connecting F_1 and F_2. The point P is the intersection of these two lines.

4. The Witch is the locus of points P generated as θ is allowed to vary.

- Cartesian:
$$y = \frac{8R^3}{x^2 + 4R^2}$$

- Parametric:
$$x = 2R\cot\theta$$
$$y = R(1 - \cos 2\theta)$$

Example: EQ: `'(2*R/TAN(θ),R*(1-COS(2*θ)))'`
TYPE: Parametric   ∡: Rad   INDEP: θ
H-VIEW: -10 10   V-VIEW: -3 7
LO: 0   HI: 3.1416 STEP: .05

$R = 2$

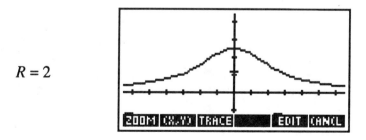

4. Polynomials

Polynomials and their Characteristics

The term *polynomial* has a more limited meaning on the HP 48 than in math text-books. Expressions such as $4xy^3 - 7x^2y - 3y$, with two or more variables, are also polynomials. But the HP 48 (and this chapter, too) limits its definition of polynomial to "polynomials in a single variable." (It *can* handle expressions with two or more variables, but it doesn't treat them as polynomials.) So a *polynomial* here is a function of the form $P(x) = a_n x^n + a_{n-1} x^{n-1} + \cdots + a_1 x + a_0$, where n is a positive integer. The real numbers, a_n, a_{n-1}, a_{n-2}, ..., a_1, a_0, are the *coefficients* of the polynomial. If $a_n \neq 0$, the polynomial is said to have *degree n*.

Note that there are two important aspects to this definition. First, a polynomial is a *function*, which means it will pass the vertical line test. Second, because it has a single variable, polynomials differ from one another only in their set of coefficients, which allows the HP 48 to compute with a polynomial more rapidly than with many other functions by using a *vector* of its coefficients. For example, $2x^5 - 3x^4 + x^3 + 6x^2 - 18x + 11$ would become $[\ 2\ \text{-}3\ 1\ 6\ \text{-}18\ 11\ ]$; and $2x^5 + x^3 + 11$ would be $[\ 2\ 0\ 1\ 0\ 0\ 11\ ]$.

Note here that coefficients for missing terms are included as zeroes, to distinguish between, say, $2x^5 + x^3 + 11$ and $2x^2 + x + 11$. Note, too, that although *computations* with polynomials are faster in vector form, you must still use their standard algebraic form to *plot* them—but you can use the polynomial solving application to help you convert from vector to symbolic form before plotting.

Example: Using the `Solve poly...` application, enter the polynomial $[\ \text{-}5\ \text{-}3\ 3\ 2\ 0\ 1\ ]$ and convert it to its symbolic form.

1. From the stack, open the `Solve poly...` application: `→`SOLVE `▼``▼` ENTER.

2. With the **COEFFICIENTS [AN ... A1 A0]** field highlighted, enter the polynomial: `←``{}``5``+/-``SPC``3``+/-` `SPC``3``SPC``2``SPC``0` `SPC``1` ENTER.

3. Re-highlight the **COEFFICIENTS** field (`▲`), and then press ▐SYMB▐ CANCEL to see the symbolic form on stack level 1:

$$'\text{-}(5*X^5)\text{-}3*X^4+3*X^3+2*X^2+1'$$

Graphs of Polynomials

The graph of a polynomial tells a lot about it. You can find the number of real roots, estimate the degree and also points of local maxima and minima.

Example: Plot the polynomial created in the previous example.

1. Open the **PLOT** application (⟨→⟩⟨PLOT⟩) and change the plot type to Function (⟨▲⟩⟨α⟩⟨F⟩).

2. Highlight the **EQ:** field and grab the polynomial from stack level 1: ⟨▼⟩ ⟨NXT⟩ **CALC** (⟨◀⟩, if necessary to put the target polynomial in level1) **OK**.

3. Set **H-VIEW** to ⁻2 3 and **V-VIEW** to ⁻3.5 4.5.

4. Set **INDEP:** to X and the plotting range and step size (in the **PLOT OPTIONS** screen) to their defaults (**RESET** each field, if necessary).

5. Plot the function (**OK ERASE DRAW**).

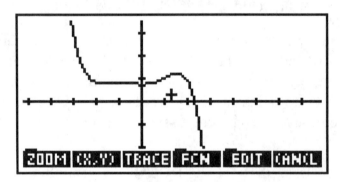

Observe the plot to see what you can determine from it.

- The plot of a polynomial will cross the x-axis once for each real root. This polynomial appears to have only one real root.

- The plot of a polynomial will have one fewer "bend" than its degree. However, the number of "bends" is not always immediately obvious. This plot *appears* to have two bends—like a third-degree polynomial—but you know from its equation that it is in fact a fifth-degree polynomial. A plot's extra bends may be "hiding" within one of the visible bends that doesn't curve very sharply, but which is rather flat and broad.

Look again at the plot and notice that the left-most bend looks like it might be a suspect for such hidden bends. Test your suspicion by using the Box-zoom to magnify the flat region: Move the cursor to the upper-left corner of the region you want to magnify and press ▮ZOOM▮ ▮BOX▮ . Then press the ▶ and ▼ keys until the zoom-box encloses the flat region that you're investigating:

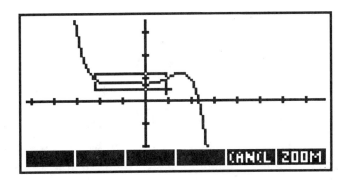

Press ▮ZOOM▮ to draw the magnified region:

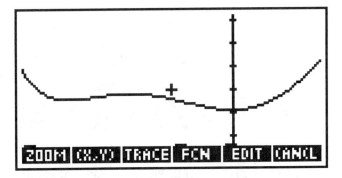

As you can see, the "flat" bend is actually composed of three bends; the polynomial has four bends altogether (which is, after all, what you would expect of a fifth-degree polynomial).

Actually, each "bend" in a polynomial represents spot where the *slope* of the polynomial "levels out"—to zero. If you plot the *slope* of the polynomial (known as the *first derivative* of the polynomial) instead of the polynomial itself, you can count the number of times this plot crosses the *x*-axis (i.e. the number of times that the slope is zero) to determine the number of "bends" in the original polynomial.

Example: Plot the previous polynomial again using the original display coordinates. Then plot the first derivative of the polynomial.

1. Press [CANCEL] to return to the **PLOT** screen.

2. Reset **H-VIEW** to ⁻2 3 and **V-VIEW** to ⁻3.5 4.5.

3. Redraw the polynomial: **ERASE** **DRAW**.

4. Draw the first derivative of the polynomial: **FCN** [NXT] **F'** . Both the polynomial and its derivative will be drawn:

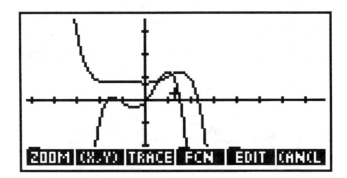

Clearly, the graph of the first derivative (slope) crosses the *x*-axis (i.e. becomes zero) in four spots—four real roots—so the original polynomial is a fifth-degree polynomial.

The plot of the first derivative can also be used to determine more precisely the location of the bends—which are local maximums and minimums (*extrema*, to your HP 48). The *x*-coordinate of the extremum is the same as the *x*-coordinate at the corresponding *zero* of the first derivative.

Example: Find the coordinates of the left-most extremum of the polynomial by finding the corresponding zero of its first derivative.

1. While viewing the plot of the polynomial and its first derivative, move the cursor so that it's close to the left-most zero of the derivative function. Remember: whenever there are two or more functions plotted, only one of them is the *current* function. The derivative function is the current function at the moment.

2. Press **FCN** **ROOT**. Note that the cursor moves to the actual spot of the root being solved. Make sure that it is the one you intended— this is a good check to be sure you've communicated properly with your HP 48. Result: **ROOT: -0.657387549433**

3. The *x*-coordinate of the left-most extremum is -0.66. To find the *y*-coordinate, switch the current function to the polynomial ([NXT] [NXT] **NXEQ**), and then compute the value of the polynomial at $x \approx -0.66$ ([NXT] **F(X)**). Result: **F(X): 1.06564988769**

Thus, the coordinate of the left-most extremum is about (-0.66, 1.07).

The HP 48, of course, can find the coordinates of an extremum more directly if it is easily distinguished from others. The method of the previous example is usually better when extrema ("bends") are hidden or very close together, but another method is quicker when the extremum is easy to "point out."

Example: Find the coordinates of the right-most extremum directly from the plot of the polynomial itself.

1. Assuming that the original polynomial is still the current function, move the cursor right to a point near the right-most extremum.

2. Press [NXT] (to redisplay the menu) [NXT] **EXTR** to compute the nearest extremum. Result (to 2 places): **EXTRM: (0.59, 1.59)**

The previous two examples may suggest that every polynomial has exactly one less "bend" (extremum) than its degree. But that's not true, particularly in polynomials with some coefficients equal to zero. The next example shows how you can determine the degree from the plot of these exceptional polynomials.

Example: Plot the polynomial, [-5 0 3 2 0 1], and its first derivative. Demonstrate graphically the degree of the polynomial.

1. Return to the stack ([CANCEL][CANCEL]) and move to the Solve poly... application ([→][SOLVE][▼][▼][ENTER]).

2. Enter the polynomial into the **COEFFICIENTS** field: [←][{ }][5][+/−] [SPC][0][SPC][3][SPC][2][SPC][0][SPC][1][SPC][ENTER].

3. Create the symbolic version: [▲] **SYMB** [CANCEL].

4. Open the **PLOT** application, enter the symbolic polynomial, and, using the same settings as with the previous polynomial (they should still be there), plot the polynomial: [→][PLOT][NXT] **CALC** [←] **OK** [NXT] **ERASE** **DRAW**.

5. Add the first derivative to the plot: **FCN** [NXT] **F'** :

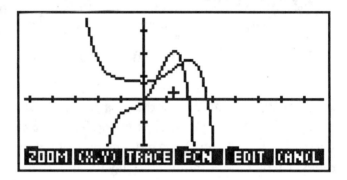

Notice that original polynomial appears to have two bends (i.e. is third-degree). And the first derivative plot seems to concur: it has two zeroes, exactly the number expected for a third-degree polynomial. But look at the shape of the first derivative: it appears to have *three* bends. A first derivative cannot have more bends than its original polynomial. In fact, the first derivative of a third-degree polynomial can have no more than one bend. This is a powerful clue that the original polynomial is actually fifth-degree, at least.

For polynomials of high degree that lack most lower-degree terms, you may have to find the derivative of the derivative (*the second derivative*), or the derivative of that (*the third derivative*), etc., until the necessary clues appear. If at any stage, the result is a straight-line, then the degree of the polynomial is one higher than the number of derivatives it took to generate the straight line.

Example: Continue plotting higher derivatives for the polynomial in the previous example until the result is a straight-line.

1. Assuming that the plot of the polynomial and its first derivative is still displayed, press **FCN** (NXT) **F'** to add the plot of the second derivative to the display:

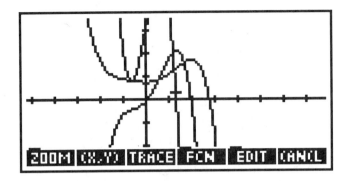

Sure enough, the second derivative still has two bends (one of them occurring abruptly at a hidden point of inflection in the original polynomial's broad left-most bend).

2. Repeat step 1 and generate the third derivative. Because the plots are getting steeper and narrower, perform some zooming to adjust the viewing scale. Press **ZOOM ZFACT** to be sure that the zoom-factor is set to the default, 4. Press **OK** (NXT) **VZOUT** (Vertical Zoom OUT), then interrupt the drawing ((CANCEL)) and press **ZOOM** (NXT) **HZIN** (Horizontal Zoom IN). Interrrupt again, move the cursor to the origin and press **ZOOM** (NXT) **CNTR**. All zooms will be reflected in the set of plots finally drawn:

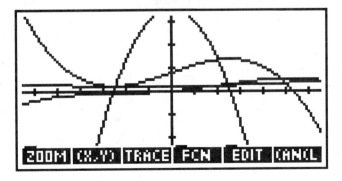

This appears to be a parabola with one bend.

3. Repeat step 1 again and generate the fourth derivative:

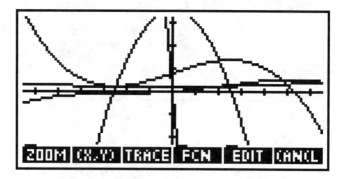

The fourth derivative appears to be a nearly vertical line. Zoom out to confirm that it isn't merely an illusion caused by the current display settings. Press **ZOOM** (NXT) **VZOUT**. Press **FCN** (NXT), then press and hold down **VIEW**. This displays the symbolic expression of the current function (the fourth derivative).

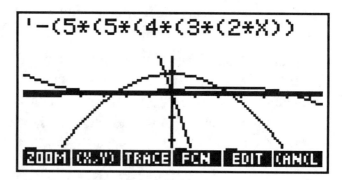

The fourth derivative is a line—a first-degree polynomial—which demonstrates that the original polynomial was a fifth-degree polynomial.

Polynomial Arithmetic

The easiest way to do arithmetic with polynomials—add, subtract, multiply, divide, and raise to a power—is to use the vector form of polynomials.

Addition and Subtraction

The addition and subtraction of polynomials is most easily accomplished by adding or subtracting corresponding coefficients in the two polynomials. For example, adding $x^4 + 3x^3 - 7x^2 - 5x + 17$ and $-5x^3 + 3x - 4$ is simply a matter of adding the coefficients of like terms:

$$
\begin{array}{r}
x^4 + 3x^3 - 7x^2 - 5x + 17 \\
+ \qquad -5x^3 \qquad\quad + 3x - 4 \\
\hline
x^4 - 2x^3 - 7x^2 - 2x + 13
\end{array}
$$

On the HP 48, you can perform polynomial addition and subtraction in either of two ways—symbolically, using the built-in algebraic abilities or "numerically" using the vector form of the polynomial and the program PADD (see page 289 for listing).

Example: Add $x^4 + 3x^3 - 7x^2 - 5x + 17$ and $-5x^3 + 3x - 4$, using the built-in symbolic tools of the HP 48.

1. From the stack, enter the first polynomial in its symbolic form: ⟨←⟩
   EQUATION α ⟨←⟩ X yˣ 4 ▶ + 3 α ⟨←⟩ X yˣ 3 ▶ - 7 α ⟨←⟩ X
   yˣ 2 ▶ - 5 α ⟨←⟩ X + 1 7 ENTER.

2. Enter the second polynomial in symbolic form: ⟨←⟩ EQUATION - 5
   α ⟨←⟩ X yˣ 3 ▶ + 3 α ⟨←⟩ X - 4 ENTER.

3. Add the polynomials; collect like terms: + ⟨←⟩ SYMBOLIC **COLCT**.
   Result: `'13-7*x^2-2*x^3+x^4-2*x'`

   It's all there, even if it is a bit out of order.

Example: Add those same two polynomials, $x^4 + 3x^3 - 7x^2 - 5x + 17$ and $-5x^3 + 3x - 4$, using their vector forms and the program PADD (assuming that you have previously entered and stored the program —see page 289—and that it is available in the current directory).

1. Enter the first polynomial: ⟵[[ ]] [1] [SPC] [3] [SPC] [7] [+/−] [SPC] [5] [+/−] [SPC] [1] [7] [ENTER].

2. Enter the second polynomial: ⟵[[ ]] [5] [+/−] [SPC] [0] [SPC] [3] [SPC] [4] [+/−] [ENTER].

3. Run the program PADD: [α][α][P][A][D][D] [ENTER].
   <u>Result:</u> [1 -2 -7 -2 13]

4. *Optional.* Now enter a variable name and use the program P→SYM (see page 296) to convert the polynomial from its vector to its symbolic form: ['] [α][X] [ENTER] [α][α][P][⟶][→][S][Y][M] [ENTER].
   <u>Result:</u> 'X^4-2*X^3-7*X^2-2*X+13'

Example: Subtract $x^4 + 3x^3 - 7x^2 - 5x + 17$ from $x^5 - 2x^2 + 12$.

1. Enter the two polynomials in the same order as you would enter two real numbers that you are subtracting: ⟵[[ ]] [1] [SPC] [0] [SPC] [0] [SPC] [2] [+/−] [SPC] [0] [SPC] [1] [2] [ENTER] ⟵[[ ]] [1] [SPC] [3] [SPC] [7] [+/−] [SPC] [5] [+/−] [SPC] [1] [7] [ENTER].

2. Perform the subtraction. You may either press [+/−], then execute PADD, or you may execute PSUB (see page 296) directly.
   <u>Result:</u> [1 -1 -3 5 5 -5]

3. Convert the result to a symbolic expression: ['] [α][X] [ENTER] P→SYM.
   <u>Result:</u> 'X^5-X^4-3*X^3+5*X^2+5*X-5'

Multiplication

The real virtues of using the vector form of polynomial are evident when you multiply two polynomials. While symbolic multiplication is technically feasible with the HP 48, it will often cost you a lot of time and patience to obtain a "legible" answer.

The program PMULT (see page 295) performs the multiplication of two polynomials in vector form.

Example: Find the product of $x^5 - 2x^2 + 12$ and $3x^3 - 4x^2 + 8x - 9$.

1. Enter the two polynomials in vector form (in either order):
   ⬅️[[ ]] [1] [SPC] [0] [SPC] [0] [SPC] [2] [+/−] [SPC] [0] [SPC] [1] [2] [ENTER] ⬅️
   [[ ]] [3] [SPC] [4] [+/−] [SPC] [8] [SPC] [9] [+/−] [ENTER].

2. Execute PMULT: [α] [α] [P] [M] [U] [L] [T] [ENTER]. You will need to view the result (press [▼]) in the Matrix Writer to see it all.

   Result: [3 -4 8 -15 8 20 -30 96 -108]

3. Convert the result to a symbolic polynomial in x: [CANCEL] ['] [α]
   ⬅️[X] [ENTER] **P→SYM**.

   Result: '3*x^8-4*x^7+8*x^6-15*x^5+8*x^4+20*
   x^3-30*x^2+96*x-108'

Division

Division of polynomials does not always result in a polynomial. The result is a quotient (a polynomial) and a remainder—a rational fraction that can't be further simplified. If the remainder is zero, then the result is a polynomial. If the remainder isn't zero, then it is the ratio of two polynomials, the denominator polynomial being of the same or higher degree than the numerator polynomial.

The program PDIVIDE (see page 291) takes two polynomials in vector form as inputs—in the same order as division of two real numbers. It returns four objects:

- the quotient polynomial (level 4);
- the numerator polynomial of the remainder (level 3);
- the denominator polynomial of the remainder (level 2);
- a complete and exact algebraic result of the division—including the remainder as a rational fraction (level 1).

Example: Divide $16x^4 + 26x^3 - 61x^2 + 16x + 3$ by $2x - 1$.

1. Enter the numerator in the division: ⏪[]16 SPC 26 SPC 6 1+/− SPC 16 SPC 3 ENTER.

2. Enter the denominator: ⏪[]2 SPC 1+/− ENTER.

3. Execute PDIVIDE: α α P D I V I D E ENTER.

   <u>Result:</u>
   ```
   4:        [ 8 17 -22 -3 ]
   3:                  [ 0 ]
   2:               [ 2 -1 ]
   1:  '8*x^3+17*x^2-22*x-3'
   ```

In this case, the result is a polynomial with no remainder:
$$8x^3 + 17x^2 - 22x - 3$$

Example: Divide the polynomial $x^5 - 2x^2 + 12$ by $2x - 1$.

1. Enter the numerator polynomial: ⏎(←)([ ])(1)(SPC)(0)(SPC)(0)(SPC)(2)
(+/−)(SPC)(0)(SPC)(1)(2)(ENTER).

2. Enter the denominator polynomial: (←)([ ])(2)(SPC)(1)(+/−)(ENTER).

3. Execute PDIVIDE: (α)(α)(P)(D)(I)(V)(I)(D)(E)(ENTER).

<u>Result:</u>
```
4:  [ .5 .25 .125 -.9375 -.46875 ]
3:                     [ 11.53125 ]
2:                          [ 2 -1 ]
1:  '1/2*x^4+1/4*x^3+1/8*x^2-15/16*
    x+(-(15/16*x)+12)/(2*x-1)'
```

Here is how to interpret the result:

- The polynomial part of the quotient is returned to level 4. In this case it represents $\dfrac{1}{2}x^4 + \dfrac{1}{4}x^3 + \dfrac{1}{8}x^2 - \dfrac{15}{16}x - \dfrac{15}{32}$.

- The numerator of the remainder is on level 3; the denominator is on level 2; so the remainder here is $\dfrac{11.53125}{2x-1}$ or $\dfrac{369}{32(2x-1)}$.

- The algebraic on level 1 is the exact result of the division and—as in this case—may not be a polynomial. Note how the algebraic incorporates the remainder into the polynomial part of the quotient. (Of course, if the remainder's numerator on level 3 is [0], the algebraic will simply be the level-4 quotient in its symbolic form—as in the previous example.)

PDIVIDE offers one approach to polynomial division. *Synthetic* division, discussed on pages 115-123, is another approach often used in finding the roots of polynomials.

Finding Positive Integral Powers of a Polynomial

The program PPOWER (see page 296) makes it a lot easier and quicker to find expansions of polynomials than by repeatedly multiplying their symbolic expressions and expanding and collecting. PPOWER considers only the positive integral powers of polynomials, so that the result is an ordinary polynomial (explicitly computing the 1/3 power or the -2 power of a polynomial usually complicates the expression a great deal without adding much new information).

Example: Find the fifth power of the polynomial $x^2 + 6x - 10$ using PPOWER (assuming that it has been previously entered and is available in the current directory).

1. Put the polynomial on the stack in vector form: ⦅←⦆⦅[ ]⦆⦅1⦆⦅SPC⦆⦅6⦆ ⦅SPC⦆⦅1⦆⦅0⦆⦅+/−⦆⦅ENTER⦆.

2. Put a *vector* containing the power on the stack: ⦅←⦆⦅[ ]⦆⦅5⦆⦅ENTER⦆.

3. Execute PPOWER: ⦅VAR⦆ ⦅NXT⦆ or ⦅←⦆⦅PREV⦆ as needed) **PPOW**. View the results by pressing ⦅▼⦆ and then ⦅▶⦆ as needed.

   Result: [1 30 310 960 −3320 −17424 33200
            96000 −310000 300000 −100000]

4. *Optional.* Convert the result to a symbolic expression: ⦅CANCEL⦆ if necessary to exit the Matrix Writer, then ⦅'⦆⦅α⦆⦅←⦆⦅X⦆⦅ENTER⦆⦅VAR⦆ (and ⦅NXT⦆ or ⦅←⦆⦅PREV⦆ as needed), then **P→SY**.

   Result: 'x^10+30*x^9+310*x^8+960*x^7− 3320*x^6−
           17424*x^5+33200*x^4+96000*x^3−
           310000*x^2+300000*x−100000'

Finding Roots of Polynomials

The *roots* of a polynomial $P(x)$ are those values of x that satisfy the equation $P(x) = 0$ (hence the other common names for roots, *zeroes*). There are several ways to find the roots of polynomials using the HP 48:

- Use synthetic division, guided by information obtained from a set of polynomial theorems, to manually search for roots.

- Use the Solver.

- Find roots graphically.

- Use root-finding algorithms customized for polynomials.

The traditional "manual" means of finding roots of polynomials required lots of trial-and-error computations involving polynomial division. To streamline these computations, the notational shortcut known as *synthetic division* was developed. Further, various theorems were used to help one narrow the search and reduce the number of computations. Look at each of these shortcuts.

Synthetic Division

Synthetic division reduces a polynomial to its coefficients (much like the vector form that you've seen earlier in this chapter). The factor being tested is also reduced to a single number. The division problem thus resembles regular long division. For example, dividing $x^3 + 4x^2 + 3x - 2$ by $x - 3$ goes like this:

$$3\overline{)\ 1\quad 4\quad 3\quad \text{-}2}$$

Bring down the first coefficient:

$$3\overline{)\ 1\quad 4\quad 3\quad \text{-}2}$$
$$1$$

Multiply the resulting coefficient (1) by the factor (3) and add the product to the next lower coefficient in the original polynomial (4):

$$3\overline{)\ 1\quad 4\quad 3\quad \text{-}2}$$
$$\underline{\qquad 3 \qquad\qquad}$$
$$1\quad 7$$

Continue likewise through all coefficients in the polynomial:

$$\begin{array}{r} 3\overline{)\,1\quad 4\quad 3\quad -2} \\ \underline{3\quad 21\quad 72} \\ 1\quad 7\quad 24\quad 70 \end{array}$$

Read the answer from the bottom line: The correct quotient is $x^2 + 7x - 24$; the final value (70) is the remainder. This shortcut works because, in fact, you *are* just

doing long division:

$$\require{enclose}\begin{array}{r} x^2 + 7x + 24 \\ x-3\enclose{longdiv}{x^3 + 4x^2 + 3x - 2} \\ \underline{x^3 - 3x^2} \\ 7x^2 + 3x \\ \underline{7x^2 - 21x} \\ 24x - 2 \\ \underline{24x - 72} \\ 70 \end{array}$$

Of course, such synthetic division is suitable for manual computation, but you can shorten its work by automating the process using your HP 48. The next two examples illustrate two different approaches.

Example: Use the SDIV program (see page 303) to do the synthetic division of the polynomial $3x^5 + 2x^4 - 7x^3 + 3x - 9$ by the factor -3.

1. Enter the vector form of the polynomial: ⟨←⟩ [ ] 3 SPC 2 SPC 7 +/− SPC 0 SPC 3 SPC 9 +/− ENTER.

2. Enter the factor: 3 +/− ENTER.

3. Execute SDIV: α α S D I V ENTER or VAR (NXT or ←PREV as needed) SDIV.

   Result: 2: [3 -7 14 -42 129]
            1:                    -396

The quotient, in vector form, is returned to level 2, the remainder to level 1. The original polynomial and factor are returned to levels 4 and 3, respectively, in case you want to use SDIV repeatedly with the same polynomial (which you often will).

4. POLYNOMIALS

Example: Repeat the previous example using the program SYND (see page 313).

1. Execute the SYND program: α α S Y N D ENTER or VAR (NXT or ← PREV as needed) **SYND**.

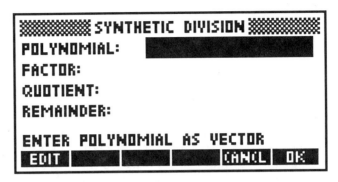

2. Enter the polynomial in the **POLYNOMIAL:** field: ← { } 3 SPC 2 SPC 7 +/− SPC 0 SPC 3 SPC 9 +/− ENTER.

3. Enter the factor in the **FACTOR:** field: 3 +/− ENTER.

4. Press **OK** to perform the synthetic division.

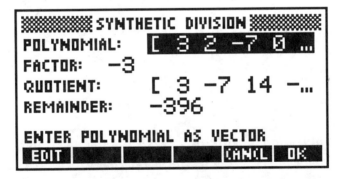

5. The SYND program is designed for repeated use of synthetic division, allowing you to search for roots without affecting the stack. Press CANCEL when you want to exit the program.

Polynomial Theorems

Wise use of synthetic division involves narrowing the number of factors that you must try in your search for roots. And there are a number of theorems which can help you do just that. With their proofs omitted, they are:

1. **Fundamental Theorem of Algebra:** *Every polynomial equation with degree greater than zero has at least one root in the set of complex numbers.*

2. **Corollary to Fundamental Theorem:** *A polynomial equation of degree n has exactly n roots in the set of complex numbers, where two or more roots with the same value are treated as distinct.*

   A fifth-degree polynomial, for example, has five complex roots, some of which may also be real, some of which may be identical to each other.

3. **Complex Conjugates Theorem:** *If a and b are real numbers with b ≠ 0 and the complex number a + bi is a root of a polynomial equation, then its conjugate, a − bi, is also a root of the polynomial.*

   Thus, polynomials can have only an even number of non-real complex roots. So, polynomials of odd degree *must* have at least one real root.

4. **Remainder Theorem:** *If a polynomial P(x) is divided by (x − a), then the remainder is a constant, P(a).*

5. **Factor Theorem:** *If a polynomial P(x) is divided by (x−a) and the remainder, P(a), is zero, then a is a root of P(x).*

6. **Rational Root Theorem:** *If a polynomial has rational roots of the form p/q, where p/q is in simplest form, then p is a factor of the constant term and q is a factor of the coefficient of the highest-degree term.*

   For example, if the polynomial $6x^3 - 3x^2 + x + 7$ has rational roots (p/q), then p is a factor of the constant term, 7 (i.e. either ± 1 or ± 7) and q is a factor of the coefficient of the highest-degree term, 6 (i.e. either $\pm 1, \pm 2, \pm 3,$ or ± 6). Thus, the only possible rational roots of this polynomial are:

$$\pm 1, \ \pm \frac{1}{2}, \ \pm \frac{1}{3}, \ \pm \frac{1}{6}, \ \pm 7, \ \pm \frac{7}{2}, \ \pm \frac{7}{3}, \ \pm \frac{7}{6}$$

7. **Descartes' Rule of Signs:** *If $P(x)$ is a polynomial whose terms are arranged in descending powers of the variable, then the number of positive real roots of $P(x)$ is the same as the number of changes in sign of the coefficients of the terms, or is less than this number by an even multiple; and the number of negative real roots of $P(x)$ is the same as the number of changes in the sign of $P(-x)$, or is less than this number by an even multiple.*

   For example, for the polynomial $2x^4 - x^3 + 5x^2 + 3x - 9$, the signs of the coefficients in descending order are $\{+-++-\}$. Reading from left to right, there are three changes in sign. Therefore, there are either 3 or 1 positive real roots of $P(x)$. Next, evaluate $P(-x)$ and apply the rule of signs to assess the number of negative real roots. $P(-x) = 2x^4 + x^3 + 5x^2 - 3x - 9$ and the signs are $\{+++--\}$. There's only one change and thus there is exactly one negative real root.

8. **Upper Bound Theorem:** *If c is positive and $P(x)$ is divided by $x - c$ and the resulting quotient and remainder have no change in sign, then $P(x)$ has no real roots greater than c. Thus c is an upper bound of the roots of $P(x)$.*

   For example, to test whether 4 is an upper bound of the roots of the polynomial, $x^4 - 3x^3 - 2x^2 + 3x - 5$, divide the polynomial by $x-4$. All coefficients in the quotient $(x^3 + x^2 + 2x + 11)$ and remainder (39) have the same sign, so all real roots of the polynomial must be less than 4.

9. **Lower Bound Theorem:** *If c is a nonpositive number and $P(x)$ is divided by $(x - c)$ and the quotient and remainder have alternating signs, then $P(x)$ has no real roots less than c. Thus, c is a lower bound of the roots of $P(x)$.*

   To test whether -2 is a lower bound of the polynomial $x^3 - 2x^2 + 6$, for example, divide the polynomial by $x + 2$. The coefficients in the quotient $(x^2 - 4x + 8)$ alternate in sign, and the remainder (-10) is opposite in sign from the constant term in the quotient, thus confirming that all real roots of the polynomial must greater than -2.

Searching for Roots with Synthetic Division

Now that you have been introduced to the essential tools, try a few examples.

Example: Find the roots of $P(x) = 5x^5 - 16x^4 - 7x^3 + 52x^2 - 70x + 12$.

    1. Use APOLY (see page 276) to apply Descartes' Rule of Signs and to find an integral lower bound and an integral upper bound for the set of real roots of the polynomial. Enter the polynomial in vector form onto the stack and then type α α A P O L Y ENTER.

Results: 3:   Signs: { 4 1 }

2:            Range: { -3 4 }

1: [5 -16 -7 52 -70 12]

    2. Use the Rational Roots Theorem to compile a set of possible rational roots within the range determined in step 1 (note that -3 and 4 cannot be roots because they represent bounds; roots are found _between_ them). Since $p = \{ \pm1, \pm2, \pm3, \pm4, \pm6, \pm12 \}$ and $q = \{ \pm1, \pm5 \}$, possible p/q's within the range are:

$$\left\{ -\frac{12}{5}, -2, -\frac{6}{5}, -1, -\frac{4}{5}, -\frac{3}{5}, -\frac{2}{5}, -\frac{1}{5}, \frac{1}{5}, \frac{2}{5}, \frac{3}{5}, \frac{4}{5}, 1, \frac{6}{5}, 2, \frac{12}{5}, 3 \right\}$$

    3. Since you know there is one negative real root, begin using synthetic division with the most negative of the possible rational roots, -12/5. The polynomial should still be on level 1 after the execution of APOLY. Key in the factor, 1 2 +/− ENTER 5 ÷, then VAR **SDIV**.

Result: 2: [5 -28 60.2 -92.48 151.952]

1:                          -352.6848

    Note that the quotient and remainder have alternating signs, which means that the factor is a lower bound.

    4. Press ◄ ◄ ◄ and repeat step 3 using -2 as the factor.

Result: 2: [5 -26 45 -38 6]

1:               0

    Aha! You've found the negative real root—indicated by the zero remainder.

5. You can now use the quotient from step 4 when you found an exact root. It is sometimes referred to as the *depressed polynomial*. Press ← and begin your hunt for positive roots with the most positive of the possible set of rational roots, 3: (3)(ENTER) ⬛SDIV⬛.

Result: 2:   [5 -11 12 -2]
           1:                   0

You're on a roll—you've found another root. Now that you have one positive real root, you know from Descartes' Rule of Signs that there must be either one or three more positive real roots. If you can find one more, you can use the quadratic equation to find the other two.

6. Repeat the search using other possible positive rational roots. You can either continue to test roots in descending order, or you can test a sampling and watch for remainders changing signs. Try the latter method. Press ← and repeat the SDIV process on the latest depressed polynomial ([5 -11 12 -2]) using factors of 2, 1, and 0.

Results: (2)   2:   [5 -1 10]
                  1:                  18
           (1)   2:   [5 -6 6]
                  1:                   4
           (0)   2:   [5 -11 12]
                  1:                 -2

Because the remainder changes signs between 1 and 0, you know that at least one real root lies between 0 and 1 (possibly closer to 0).

7. Review the list of possible rational root values for those between 0 and 1. Repeat the search starting with the value nearest zero, 1/5.

Result: 2:   [5 -10 10]
             1:                 0

Voilà! There's the third root. Now you can use the quadratic formula to force the remaining two roots into the open.

8. Press ←, then execute QDSOLV (see page 297), which applies the quadratic formula to a vector of coefficients of a quadratic equation.

Result: { '1+i' '1-i' }

Mission accomplished. The five roots of the polynomial are -2, 0.2, 3, 1+*i*, and 1-*i*.

Example: Find the roots of the polynomial $2x^3 - 3x^2 + 4x - 9$.

1. Enter the polynomial in vector form: ⟨←⟩⟨[ ]⟩⟨2⟩⟨SPC⟩⟨3⟩⟨+/−⟩⟨SPC⟩⟨4⟩ ⟨SPC⟩⟨9⟩⟨+/−⟩⟨ENTER⟩. Your goal is simply to find *one* root—the other two can be obtained directly from the quadratic equation.

2. Execute ⟨APOLY⟩ to analyze the polynomial: ⟨VAR⟩ ⟨APOLY⟩.

   Result: 3:   Signs:{ 3 0 }
            2:   Range:{ 0 3 }
            1:   [2 -3 4 -9]

   There are either 3 or 1 positive real roots, and no negative real roots —a fact confirmed by zero being the integral lower bound.

3. List the possible rational roots in the range. With $p = \{ \pm 1, \pm 3, \pm 9 \}$ and $q = \{ \pm 1, \pm 2 \}$, this list of candidates is short: $\left\{ \dfrac{1}{2}, 1, \dfrac{3}{2} \right\}$

4. Begin the ⟨SYND⟩ program and enter the polynomial into the **POLY-NOMIAL:** field: ⟨VAR⟩ ⟨SYND⟩, then ⟨NXT⟩ ⟨CALC⟩ ⟨OK⟩.

5. Enter the most positive candidate factor into the **FACTOR:** field and perform the synthetic division: ⟨1⟩⟨.⟩⟨5⟩⟨ENTER⟩ ⟨OK⟩.

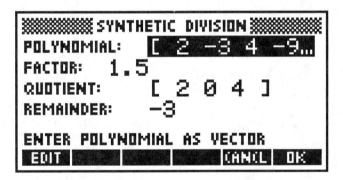

```
░░░░░░░░ SYNTHETIC DIVISION ░░░░░░░░
POLYNOMIAL:   [ 2 -3 4 -9…
FACTOR:   1.5
QUOTIENT:      [ 2 0 4 ]
REMAINDER:    -3

ENTER POLYNOMIAL AS VECTOR
 EDIT              CANCL  OK
```

The remainder is negative, unlike the positive remainder at the upper bound (3). *Conclusion*: There exists a real root between 1.5 and 3.

6. Before seeking the real, non-rational root between 1.5 and 3, you might try the other two rational root candidates in the list. (If the list were longer, it may not be worth it to do this.). Enter the factors 1 and then .5 and run the synthetic division. Note the remainders.
   Results: For 1: **REMAINDER: -6**;   for .5: **REMAINDER: -7.5**

Not only are the remainders still negative (i.e. there has been no change from 1.5), they are getting more so. Furthermore, the quotient and remainder are alternating signs, suggesting that these are lower bounds. *Conclusion*: There are no more positive real roots other than the one between 1.5 and 3 that you isolated in step 5.

7. Narrow down the range where the root lives by using the *bisection method*. In this method, you choose as your next factor the approximate midpoint of the range where you know the root to be located, and keep repeating this choice as you narrow the range. Begin with 2.25 as the factor. Here are your re<u>sults</u>:

(2.25) **REMAINDER: 7.59375**   (Positive; root is smaller)
(1.88) **REMAINDER: 1.206144**   (Positive; root is smaller)
(1.69) **REMAINDER: -1.154682**   (Negative; root is larger)
(1.785) **REMAINDER: -.04385175**   (Negative; root is larger)
(1.833) **REMAINDER: .569686074**   (Positive; root is smaller)
(1.809) **REMAINDER: .258393258**   (Positive; root is smaller)
(1.797) **REMAINDER: .106150146**   (Positive; root is smaller)
(1.791) **REMAINDER: .030870342**   (Positive; root is smaller)
(1.788) **REMAINDER: -.006560256**   (Negative; root is larger)
(1.789) **REMAINDER: .005901138**   (Positive; root is smaller)
(1.7885) **REMAINDER: -.000331491**   (Negative; root is larger)

You could continue this to the 12-digit limit of the HP 48, but 3 or 4 digits is usually enough. The approximate real root is 1.7885.

8. Use **QDSOLV** to compute the other two roots (also approximate, because the real root is approximate). Assuming you have just computed the synthetic division for 1.7885, highlight the **QUOTIENT:** field and press [NXT] **CALC** [ENTER] **OK** [CANCEL] to move a copy of the computed quotient—a quadratic—to the stack. Then execute **QDSOLV:** [VAR] **QDSO**.

Results (displayed to 4 places): { '-0.1443+1.5796*i'
                                              '-0.1443-1.5796 *i' }

Thus, the three roots of the polynomial (approximate to 4 decimal places) are: 1.7885   -0.1443+1.5796*i*   -0.1443–1.5796*i*

Using the Solver to Find Roots

As you can see, the process of "zooming" in manually on approximate real roots can be quite tedious. The built-in Solver of the HP 48 can speed up this process considerably. The next two examples use the same two polynomials as the previous two examples, but this time, the built-in Solver is used.

Example: Find the roots of $P(x) = 5x^5 - 16x^4 - 7x^3 + 52x^2 - 70x + 12$.

1.   Enter the polynomial in vector form onto the stack and make an extra copy: [EVAL] [←][[ ]] [5] [SPC] [1][6][+/−] [SPC] [7][+/−] [SPC] [5][2] [SPC] [7][0][+/−][SPC][1][2] [ENTER] [ENTER].

2.   Use **APOLY** to apply Descartes' Rule of Signs and find an integral lower bound and an integral upper bound for the set of real roots of the polynomial. Enter the polynomial in vector form onto the stack and then type [α][α][A][P][O][L][Y] [ENTER].

   <u>Result:</u>   3: Signs: { 4 1 }
   2: Range: { -3 4 }
   1: [5 -16 -7 52 -70 12]

   As before, this gives you some idea about the distribution of roots.

3.   Convert the vector form of the polynomial to the symbolic: ['][α][←][X] [ENTER] [VAR] **P→SY**.

   <u>Result:</u> '5*x^5-16*x^4-7*x^3+52*x^2-70*x+12'.

4.   Store the equation as the current equation (in the variable EQ) and begin the Solver: [←][SOLVE] **ROOT** [←] **EQ** **SOLVR**.

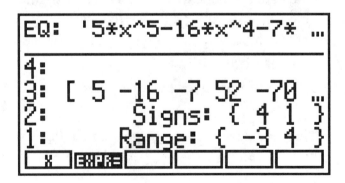

5. To search for the negative real root, enter, as a list, the range of values that bracket it; put the list in x: (←)[{}] [3] [+/−] [SPC] [0] [ENTER] [X].

6. Launch the root-finder. It will start with the range you specified and stop when it comes to a root: (←) [X]. Result: X: -2.

7. It found the negative real root. Enter the range for the positive real roots and start the root-finder again: (←)[{}] [0] [SPC] [4] [ENTER] [X] (←) [X]. Result: X: .2.

8. It found one positive real root, but Descartes' Rule of Signs says that there must be another. To find it, just specify a different starting search range than that of the root you just found. You already found a root in the first half of the range, so try the second half, { 2 4 }: (←)[{}] [2] [SPC] [4] [ENTER] [X] (←) [X]. Result: X: 3.

9. Now that you've found the three real roots, you need to return to the vector approach to polynomials; the Solver *cannot* find complex or imaginary solutions.

   Press (←) repeatedly until the vector form of the original polynomial is on level 1. Then use SDIV to "depress" the polynomial to a quadratic by "removing" the roots you just found: [2] [+/−] [VAR] SDIV (←) [.] [2] SDIV (←) [3] SDIV (←).

10. Use QDSOLV to find the remaining roots: QDSO.
    Result: { '1+i' '1-i' }

Example: Find the roots of the polynomial $2x^3 - 3x^2 + 4x - 9$.

1. Enter the polynomial in vector form: ←[ ]2 SPC 3 +/− SPC 4 SPC 9 +/− ENTER. Note that your goal is simply to find *one* root; the other two you can obtain directly from the quadratic equation.

2. Execute APOLY to analyze the polynomial: VAR APOLY.

   Result:
   ```
   3:     Signs:{ 3 0 }
   2:     Range:{ 0 3 }
   1:     [ 2 -3 4 -9 ]
   ```

3. This time, for illustration purposes, use the other Solver (you can then use either one you wish for you own work). Press →SOLVE ENTER to begin the **SOLVE EQUATION** application. The equation from the previous example will be highlighted in the **EQ:** field.

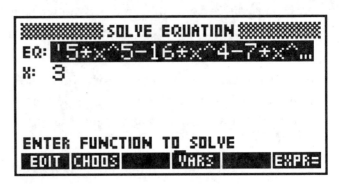

4. Use the **CALC** feature to bring a copy of the *symbolic* polynomial into the **EQ:** field: NXT **CALC** ← ENTER ' α ← X ENTER VAR **P→SY** ← CONT **OK**.

5. Move the highlight to the **X:** field and enter the range for locating a positive real root: ←[ ]0 SPC 3 ENTER. Then move the highlight to the **X:** field again and begin the root-finder: ▼ **SOLVE**.
   Result: **X: 1.78852660106**. Quick, isn't it?

6. Press CANCEL. On the stack you see the remaining copy of the polynomial vector (level 2) and the (tagged) root you just found (level 1).

7. Press VAR **SOLV** ← **QROD** to find (approximations for) the other roots. Result:
   ```
   { '-.14426330053+1.5796283111*i'
     '-.14426330053-1.5796283111*i'}
   ```

Finding Roots of Polynomials Graphically

Plotting a polynomial before trying to find its roots will give you a reasonably good idea about the location of real roots. Furthermore, the **FCN** menu, available while you are viewing the plot gives you access to the same root-finder used by the Solver—it's the best of both worlds. Look at an example.

Example: Find the real roots of $P(x) = x^5 - 6x^4 - 7x^3 + 12x^2 - 10x + 24$ graphically.

1. Begin the **PLOT** application and make sure that the **TYPE:** field contains Funct i on: →PLOT▲αF.

2. Highlight the **EQ:** field and enter the polynomial: ←EQUATION α ←X y^x 5 ▶ − 6 α ←X y^x 4 ▶ − 7 α ←X y^x 3 ▶ + 1 2 α ←X y^x 2 ▶ − 1 0 α ←X + 2 4 ENTER.

3. Enter X (lower-case) into **INDEP:**.

4. Set **H-VIEW** to −10 10 and **V-VIEW** to −5 5.

5. Press **OPTS** and make sure that the plotting range (**HI:** and **LO:**) and **STEP:** intervals are set to the default values.

6. Press **OK** **ERASE** **DRAW**:

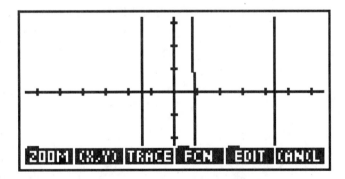

A real root exists at each intersection of the graph with the *x*-axis. In this view, you see three very clear spots where the graph crosses the *x*-axis, which is very helpful for finding roots even if it gives you a poor view of the overall polynomial.

7. Move the cursor (press ◄ several times) so that it is to the left of the most negative root. Then use the root-finder to find that nearby root: **FCN** **ROOT**. Result: **ROOT: -2.13024894272**.

8. Move the cursor near to the middle root and press ⊟ **ROOT** again. Result: **ROOT: 1.34487026015**

9. Move the cursor to the right of the most positive root and press ⊟ **ROOT**. Result: **ROOT: 6.79120116627**.

10. You've found three real roots, but are there more? Zoom out to expand your view of the polynomial: ⊟ (NXT) **PICT** **ZOOM** (NXT) **YZOUT** **ZOOM** (NXT) **YZOUT**.

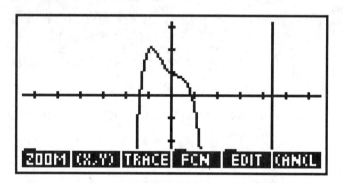

Although you still can't see all of the polynomial, you can now see enough to convince yourself that you've found all of the real roots. The "wiggle" in the graph suggests a pair of complex roots to bring the total to the required five.

Using the Built-In Polynomial Root-Finder

By far the quickest and easiest method of finding the roots of a polynomial on the HP 48 is to use its built-in *polynomial* root-finder. It uses a more specialized algorithm than the general root-finder used by the Solver.

Example: Use the built-in polynomial root-finder to find all roots of
$$P(x) = x^5 - 6x^4 - 7x^3 + 12x^2 - 10x + 24$$

1. From the stack, FIX the display to 4, then open the **SOLVE POLY-NOMIAL** application: [4] [←] [MODES] **FMT** **FIX** [→] [SOLVE] [▼] [▼] [ENTER]:

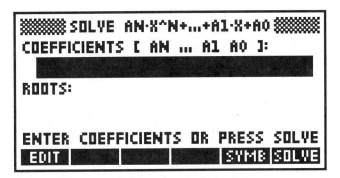

2. Enter the polynomial in vector form into the field labeled **COEFFI-CIENTS [AN ... A1 A0]:** [←] [[ ]] [1] [SPC] [6] [+/−] [SPC] [7] [+/−] [SPC] [1] [2] [SPC] [1] [0] [+/−] [SPC] [2] [4] [ENTER].

3. Press **SOLVE**.

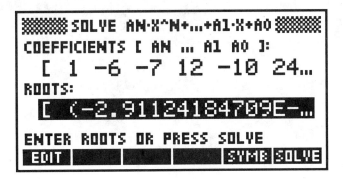

4. The result is a complex vector containing five roots expressed as complex numbers. Press ▇▇▇ to view the vector in the Matrix Writer. Then press ▶ as needed to bring each element (i.e. root) in the result vector into the command line at the bottom of the screen. The five roots, to four places, are:

$$(-0.0029, -1.1106)$$
$$(-0.0029, 1.1106)$$
$$(1.3449, 0.0000)$$
$$(-2.1302, 0.0000)$$
$$(6.7912, 0.0000)$$

The three roots whose imaginary coefficients are zero are, of course, real roots, but because the result vector contains some complex elements, the rules of HP 48 vectors require that all elements in the vector be expressed as complex numbers.

Example: Find all roots of $P(x) = x^6 + 3x^5 - 4x^4 + 10x^2 - 34x + 42$, but this time try the program **RROOTS** (see page 301), which makes use of the polynomial root-finder but provides more convenient output— a list showing real roots first.

1. Return to the stack (use [CANCEL] as needed) and enter the polynomial in vector form (be careful to note the missing third-degree term): ⬅[ ][ ]1[SPC]3[SPC]4[+/–][SPC]0[SPC]1[0][SPC]3[4][+/–][SPC] 4[2][ENTER].

2. Execute **RROOTS**: Type [α][α]R R O O T S[ENTER] or press [VAR] (then [NXT] or ⬅[PREV] as needed) ▇▇▇▇▇. <u>Result</u> (to 4 places):

$$\{ \ -3.6305 \ -2.4045 \ (1.2904, -0.5445)$$
$$(1.2904, 0.5445) \ (0.2271, -1.5496)$$
$$(0.2271, \ 1.5496) \ \}$$

Converting to Polynomials

There are many ways to find a set of roots of a given polynomial. But how would you find a polynomial that corresponds to a given set of roots? The HP 48 has a built-in function, PCOEF, that makes it easy to do just that.

Example: Find the polynomial that has the following set of roots:

$$\left\{ -\frac{1}{2}, \frac{2}{3}, \frac{7}{3}, 4+3i, 4-3i \right\}$$

1. Open the SOLVE POLYNOMIAL application: →SOLVE ▼ ▼ ENTER.

2. Highlight the ROOTS: field and enter the roots as a vector. Note that because two of the roots are complex, *all roots* must be entered as complex numbers; the real roots must use a zero as their imaginary part (and for this reason it is generally easier to enter each root onto the stack and assemble the vector at the end): ▼ NXT CALC ←() 4 ←, 3 ENTER ←() 4 ←, 3 +/- ENTER . 5 +/- ENTER 2 ENTER 3 ÷ 7 ENTER 3 ÷ 5 PRG TYPE →ARR ← CONT OK.

3. Press ▲ NXT SOLVE. This computes the polynomial's coefficients and returns it to the COEFFICIENTS field and places a copy of it on the stack. Press SYMB. This transforms the polynomial in vector form to its symbolic form and places it on the stack. Press CANCEL to view the results:

   Result: 'X^5-10.5000*X^4+45.0556*X^3
   -62.1667*X^2-4.8333*X+19.4444'

4. Since you have some rational coefficients, press ← SYMBOLIC NXT →Q.

   Result: 'X^5-21/2*X^4+811/18*X^3
   -373/6*X^2-29/6*X+175/9'

The result of the previous example is just one of the possible polynomials that have the given roots—the one whose highest-degree coefficient is 1. But if you multiply all coefficients by the least common multiple of all the rational denominators (18 here), you will get a polynomial with integral coefficients:

$$18x^5 - 189x^4 + 811x^3 - 1119x^2 - 87x + 350$$

Happily, there's a program that will save you even this step....

Example: Repeat the previous example, but this time use RCOEF (see page 298), a·program which will, given a list of roots, return a symbolic polynomial with integral coefficients.

1. Enter the roots in a *list*: ⟵ { } ' 1 +/− ÷ 2 ▶ ' 2 ÷ 3 ▶ ' 7 ÷ 3 ▶ ⟵ ( ) 4 ⟵ ' 3 ▶ ⟵ ( ) 4 ⟵ ' 3 +/− ENTER.

2. Execute RCOEF : α α R C O E F ENTER or VAR (then NXT or ⟵ PREV as needed) **RCOE**.

Result: '18*X^5−189*X^4+811*X^3
−1119*X^2−87*X+350

Another kind of conversion problem occurs when you have a function of one variable that must be converted to a polynomial before you can find its roots. The PCONV program (see page 290) is designed to perform such conversions.

Example: Find all roots of $4(x+2)^3 - \dfrac{(x-1)^2}{x+1} + 5x - 7(x-4) + 15$.

1. Enter the function in its symbolic form: ⬅️[EQUATION] [4] ⬅️[()] [α] ⬅️[x] [+] [2] ▶️ [yˣ] [3] ▶️ [−] [▲] ⬅️[()] [α] ⬅️[x] [−] [1] ▶️ [yˣ] [2] ▶️ ▶️ [α] ⬅️[x] [+] [1] ▶️ [+] [5] [α] ⬅️[x] [−] [7] ⬅️[()] [α] ⬅️[x] [−] [4] ▶️ [+] [1] [5] [ENTER].

2. Use PCONV to convert the symbolic function to a polynomial, if possible: [α] [α] [P] [C] [O] [N] [V] [ENTER] or [VAR] (then [NXT] or ⬅️[PREV] as needed) **PCON**.

   <u>Result:</u> 2: [4 28 69 123 74]
              1:               [1 1]

   The level-2 polynomial is the numerator; the level-1 polynomial is the denominator for the converted expression (so if the denominator is [1], then the conversion result is a true polynomial). Save a copy: ⬅️[STACK] [NXT] **DUP2**.

3. In any case, the roots of the original function are the same as the roots of the converted *numerator*, so press ⬅️[VAR] **RROOT** to compute the roots.

   <u>Result</u> (to 4 places): { −4.4740 −0.9256
                                 (−0.8002, −1.9563)
                                 (−0.8002, 1.9563) }

4. *Optional.* Drop the previous result and compute the actual symbolic version of the converted expression: ⬅️ **POLYN**.

   <u>Result:</u> 4:   [4 24 45 78]
                  3:                  [−4]
                  2:                  [1 1]
                  1: '4*x^3+24*x^2+45*x
                       +(78*x+74)/(x+1)'

5. SYSTEMS OF LINEAR EQUATIONS

Characterizing Systems

Systems of linear equations (and inequalities) are a powerful means of modeling real-world problems and solutions. Such systems are classified according to two characteristics:

- The relationship of the individual equations to each other.
- The ratio of independent variables to independent equations.

Any two linear equations within a system—representing lines—may either *intersect*, *not intersect*, or *coincide*. When they intersect, the equations are considered to be *consistent* and *independent*. When they don't intersect, they are *inconsistent*. When they coincide, they are *consistent* and *dependent*. When working with a system of linear equations, your goal is to include only equations that are consistent and independent with all of the others in the system, because only those equations will contribute information useful in determining a solution.

However, many kinds of real-world problems present two (or more) equations that don't precisely coincide, but are close enough to each other to *practically coincide*. Equations which coincide—exactly or practically—are called *degenerate*, which, if they are a part of a linear system that you're solving, can lead to untrustworthy solutions.

The other important characteristic of a linear system is the ratio of independent variables to independent equations. Their true *independence* is critical. Variables can appear independent—for example, test performance and shoe size—when there is actually some dependency relationship between them—age. Thus equations can appear independent (i.e. "intersecting") but be actually quite degenerate. Moral: Before deciding on the ratio of variables to equations, be sure that everything you count meets the test of independency.

If the true independence ratio is exactly 1, then the linear system has a single, exact solution. If the ratio is less than 1 (fewer variables than equations), then the linear system is *over-determined*, and you probably must settle for a *best* solution rather than an exact one. If the ratio is greater than 1 (more variables than equations), then the linear system is *under-determined*, and has either no solutions or an infinite number of them—thus requiring you to decide which is the *best* solution.

There are several ways to test the nature of a linear system: You can plot the equations together and visually check for near-coincidal lines; you can enter the system as a matrix of coefficients and find the *condition number* of the matrix; or, you can compute the *rank* of the matrix of coefficients. Look at each technique.

Example: Plot the following system of linear equations and look for one or more near-coincidal lines:

$$18x + 24y = 54$$
$$27x + 36y = 80$$
$$5x - 8y = -7$$

1. Because the **PLOT** application works best if there is no equal sign in the expressions, mentally convert each of the equations into expressions equal to zero (i.e. $18x + 24y = 54$ becomes $18x + 24y - 54$).

2. Open the **PLOT** application, make sure that **TYPE:** field says **Function**, and then reset the plot: DEL ▼ ENTER.

3. Highlight the **EQ:** field, then enter the lines in a list: ← { } ' 1 8 X α ← X + 2 4 X α ← Y − 5 4 ▶ ' 2 7 X α ← X + 3 6 X α ← Y − 8 0 ▶ ' 5 X α ← X − 8 X α ← Y + 7 ENTER.

4. Because **y** is on the same side of the equal sign as the independent variable **x**, you must be sure that some value is stored in **y** before attempting a Function plot. The value you choose will affect the location of the plots, but not the positions of each with respect to the others. Store 1 in **y**: NXT **CALC** 1 ' α ← Y STO **OK** NXT.

5. Enter **x** (lower-case) in **INDEP:**, set **H-VIEW** to −10 10; and set **Y-VIEW** to −4 6. Then press **ERASE DRAW**.

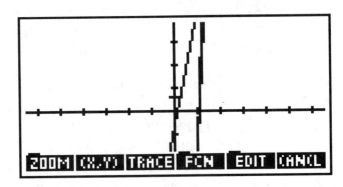

6. The two lines on the right are suspiciously close. Press **TRACE** to begin trace mode. Then press ▶ until the cursor is visible along one of the suspect lines. With the cursor on one of the lines (press ▲ as needed to switch lines) press ←, then press and hold down ▼. You will see the equation of the current line displayed above the plot until you release ▼. Repeat the procedure with the other suspect line.

Of course, visually inspecting the plots of system of linear equations is not possible for systems involving more than two independent variables.

Example: Find the condition number for the square (4 x 4) matrix representing

this linear system:
$$18x + 24y + 6z - 54$$
$$27x + 36y + 9z - 80$$
$$5x - 8y + 4z + 7$$
$$-3x + 6y - 9z + 12$$

1. Press CANCEL CANCEL to return to the stack. Then enter the matrix of coefficients for this system: →MATRIX 1 8 ENTER 2 4 ENTER 6 ENTER 5 4 +/– ENTER ▼ 2 7 ENTER 3 6 ENTER 9 ENTER 8 0 +/– ENTER 5 ENTER 8 +/– ENTER 4 ENTER 7 ENTER 3 +/– ENTER 6 ENTER 9 +/– ENTER 1 2 ENTER ENTER.

2. Make a copy of the matrix for later and find the condition number: ENTER MTH **MATR NORM COND**. Result: 1443.73404255

   The larger the condition number of a matrix, the more likely it is to be *ill-conditioned*—that is, to contain dependent equations. But how large is too large? That depends upon how many digits in your answer you want to trust.

3. Do a small computation to give a rough idea of how many digits you can trust in a solution computed using this matrix. Enter 15 (you'd use 12 if the coefficients in the matrix were themselves the result of HP 48 computations). Then press SWAP →LOG –.

   Result: 11.840512803

   Following this rule-of-thumb, you can trust up to 12 digits of any solution computed using this matrix. Thus, although they are close, the first two equations in the system are indeed independent.

However, the condition number can be found only for *square matrices*. To characterize non-square matrices, you must find their *rank*.

Example: Find the rank of the matrix representing the following system of linear equations, and use it to determine whether all its equations are independent:

$$18x + 24y + 6z - 54$$
$$27x + 36y + 9z - 80$$
$$5x - 8y + 4z + 7$$

1. You should still have a copy of the previous matrix on level 2. If you don't, enter the matrix above directly. If you do, edit it by removing the last (i.e. fourth) row: ⬅ 4 MTH **MATR** **ROW** **ROW-** ⬅.

2. Make another copy of the revised matrix and then compute the rank ENTER MTH **MATR** **NORM** NXT **RANK**. <u>Result</u>: 3

   The result indicates that all three equations are independent—the matrix is of *full rank*.

3. Press ⬅ ⬇ ⬇ ▶ ▶ ▶ 8 1 +/- ENTER ENTER to edit the matrix by changing the constant in the second equation from -80 to -81. Find the rank of the edited matrix: **RANK**. <u>Result</u>: 2

   The **RANK** computation is indicating that the second equation is now linearly dependent upon the first, so now there are only two truly independent equations.

Introduction to Matrix Arithmetic

Solving systems of linear equations of more than two equations in two variables requires the use of matrices. The HP 48 handles all manner of matrix math with ease. The following few examples illustrate some of the basic matrix operations.

Matrix Addition

Two matrices may be summed only if they have exactly the same dimensions. Matrix addition simply sums corresponding elements of the two matrices and is therefore commutative: if A and B have identical dimensions, $A + B = B + A$.

Example: Add $A = \begin{bmatrix} -2 & 4 & 2 \\ 5 & -1 & -6 \end{bmatrix}$ and $B = \begin{bmatrix} 3 & -4 & -8 \\ 0 & 5 & 2 \end{bmatrix}$

    1. Enter A onto the stack: →MATRIX 2 +/− ENTER 4 ENTER 2 ENTER ▼ 5 ENTER 1 +/− ENTER 6 +/− ENTER ENTER.

    2. Enter B onto the stack: →MATRIX 3 ENTER 4 +/− ENTER 8 +/− ENTER ▼ 0 ENTER 5 ENTER 2 ENTER ENTER.

    3. Press +. Result: `[[ 1 0 -6 ]`
                            `[ 5 4 -4 ]]`

Scalar Multiplication

Scalar multiplication is the multiplication of a matrix and a number (known as a *scalar* in this context). Each element of the matrix is multiplied by the scalar, resulting in the *scalar product*. Scalar multiplication is commutative: $n \cdot A = A \cdot n$.

Example: Find the scalar product of 5 and $A = \begin{bmatrix} -2 & 4 & 2 \\ 5 & -1 & -6 \end{bmatrix}$

    1. Enter A on the stack: →MATRIX 2 +/− ENTER 4 ENTER 2 ENTER ▼ 5 ENTER 1 +/− ENTER 6 +/− ENTER ENTER.

    2. Enter the scalar: 5 ENTER.

    3. Press ×. Result: `[[ -10 20 10 ]`
                           `[ 25 -5 -30 ]]`

Matrix Subtraction

Matrix subtraction is analogous to the subtraction of real numbers. Therefore, on the HP 48, you can do matrix subtraction (**A–B**) in any one of three ways:

1. Enter **A**, enter **B**, press ⊟.
2. Enter **B**, press +/−, enter **A**, press ⊞.
3. Enter **B**, enter -1, press ⊠, enter **A**, press ⊞.

Matrix Multiplication

Matrix multiplication (**A · B**) is defined only for two matrices that are *dimensionally compatible in the given order*: The number of columns in the first matrix (**A**) must equal the number of rows in the second matrix (**B**). The result of matrix multiplication will have the same number of rows as **A** and the same number of columns as **B**.

This table illustrates the rules for dimensional compatibility with a few examples (note that the dimensions are always listed *rows* x *columns*):

A	B	A · B
③x**2** — compare (=) — **2**x5		③x5
③x**2** — compare (≠) — **5**x2		not defined
②x**3** — compare (=) — **3**x2		②x2
①x**4** — compare (=) — **4**x1		①x1
④x**1** — compare (=) — **1**x4		④x4

Matrix multiplication combines each row of **A** with each column of **B**, in a process known as the *inner product* or *dot product*, which multiplies corresponding elements (i.e. the first element of the row by the first element of the column, etc.), then sums all the products.

Thus each row/column combination results in a single element in the result matrix. The following figure demonstrates this process for the multiplication of a 2 x 3 matrix (**A**) with a 3 x 2 matrix (**B**):

$$\begin{bmatrix} x & y & z \\ r & s & t \end{bmatrix} \bullet \begin{bmatrix} a & d \\ b & e \\ c & f \end{bmatrix} = \begin{bmatrix} xa + yb + zc & xd + ye + zf \\ ra + sb + tc & rd + se + tf \end{bmatrix}$$

Example: Find **AB** if $\mathbf{A} = \begin{bmatrix} -2 & 4 & 2 \\ 5 & -1 & -6 \end{bmatrix}$ and $\mathbf{B} = \begin{bmatrix} -3 & 5 \\ -1 & 4 \\ 7 & -2 \end{bmatrix}$.

1. Enter **A**: →MATRIX 2 +/− ENTER 4 ENTER 2 ENTER ▼ 5 ENTER 1 +/− ENTER 6 +/− ENTER ENTER.

2. Enter **B**: →MATRIX 3 +/− ENTER 5 ENTER ▼ 1 +/− ENTER 4 ENTER 7 ENTER 2 +/− ENTER ENTER.

3. Press ×. <u>Result</u>: `[[ 16 2 ]`
`[ -56 33 ]]`

Matrix Transposition

Transposing a matrix converts its rows into columns and its columns into rows—an important operation in many different matrix applications.

Example: Transpose the matrix $\mathbf{A} = \begin{bmatrix} -2 & 4 & 2 \\ 5 & -1 & -6 \end{bmatrix}$

1. Enter **A**: →MATRIX 2 +/− ENTER 4 ENTER 2 ENTER ▼ 5 ENTER 1 +/− ENTER 6 +/− ENTER ENTER.

2. Transpose it: MTH **MATR MAKE TRN** .

<u>Result</u>: `[[ -2 5 ]`
`[ 4 -1 ]`
`[ 2 -6 ]]`

Finding the Determinant of a Square Matrix

The *determinant* of a matrix is a number computed from the elements of a square matrix. It isn't defined for non-square matrices. The determinant plays a key role in determining whether a matrix has an inverse—which, in turn, is the key operation in solving a system of equations.

Example: Find the determinant of the matrix $A = \begin{bmatrix} 3 & -4 & 1 \\ 2 & 6 & 0 \\ 4 & -1 & -2 \end{bmatrix}$

1. Enter A: →MATRIX 3 ENTER 4 +/− ENTER 1 ENTER ▼ 2 ENTER 6 ENTER 0 ENTER 4 ENTER 1 +/− ENTER 2 +/− ENTER ENTER.

2. Compute the determinant: MTH MATR NORM NXT DET.

Result: −78

Matrix Inversion

There is no matrix "division"—only multiplication of one matrix with the *inverse* of another. And not all matrices have inverses; matrix inversion is defined only for some (not all) *square* matrices. The inverse of a square matrix A is the matrix A^{-1} such that $A \cdot A^{-1} = A^{-1} \cdot A = I$, where I is the *identity* matrix with the same dimensions as A. An identity matrix is a square matrix whose diagonal elements are 1 and all others are 0. For example, the 3 x 3 identity matrix is $\begin{bmatrix} 1 & 0 & 0 \\ 0 & 1 & 0 \\ 0 & 0 & 1 \end{bmatrix}$

Example: Invert the matrix $A = \begin{bmatrix} 3 & -4 & 1 \\ 2 & 6 & 0 \\ 4 & -1 & -2 \end{bmatrix}$ and compute $A \cdot A^{-1}$ to check.

1. Enter A: →MATRIX 3 ENTER 4 +/− ENTER 1 ENTER ▼ 2 ENTER 6 ENTER 0 ENTER 4 ENTER 1 +/− ENTER 2 +/− ENTER ENTER.

2. Make a copy of A and compute its inverse: ENTER 1/x. Results (shown to 5 places):
```
[[  0.15385  0.11538  0.07692 ]
 [ -0.05128  0.12821 -0.02564 ]
 [  0.33333  0.16667 -0.33333 ]]
```

3. *Optional.* Make a copy of the inverse and use the program A→Q (see page 277) to convert the array elements to fractions to see if you can obtain an exact answer: ENTER α α A → → Q ENTER.

   Results:
```
{ { '2/13'   '3/26'   '1/13'  }
  { '-(2/39)'  '5/39'  '-(1/39)' }
  { '1/3'   '1/6'   '-(1/3)'  } }
```

   Note that the result is a list of lists representing rows of the matrix. This notation is standard for representing *symbolic arrays* on the HP 48 (actual arrays don't allow algebraic objects).*

4. Check the results by multiplying A by its computed inverse (if you performed step 3, first press ◀): ×.

*Converting an array of decimal approximations to a symbolic array of fractional equivalents will not be useful if the array is the result of an inversion (or other computation) of another array with approximate decimal elements.

Solving a Linear System

There are several approaches to solving a linear system of equations. The HP 48 can assist you with any of them.

- **Gaussian Elimination.** This approach uses elementary row operations of matrices to transform the matrix representing a linear system into one from which the solution can be easily computed. This is the most commonly used approach for manual solving.

- **Cramer's Rule.** Cramer's Rule is a theorem that allows you to compute the solution of a linear system by dividing its matrix into a set of smaller ones and then using ratios of the determinants of these smaller matrices to compute the solution.

- **Matrix Inversion.** While technically both of the preceding methods implicitly use matrix inversion, there are other methods better suited to electronic computation that will efficiently generate a solution directly from the matrix representing the linear system.

These methods are described in each of the next three sections.

Solving by Gaussian Elimination

The process of *Gaussian elimination* is a common approach to solving systems of linear equations when doing them manually. It transforms the *augmented* matrix representing the linear system into an equivalent *row echelon* or *reduced row echelon* matrix, from which the solution can be easily computed.*

To understand the goal more clearly, look at examples of augmented, row echelon and reduced row echelon matrices:

Augmented:	Row Echelon :	Reduced Row Echelon:
$\begin{bmatrix} x_1 & y_1 & z_1 & \vdots & k_1 \\ x_2 & y_2 & z_2 & \vdots & k_2 \\ x_3 & y_3 & z_3 & \vdots & k_3 \end{bmatrix}$	$\begin{bmatrix} 1 & a_{12} & a_{13} & \vdots & k_1 \\ 0 & 1 & a_{23} & \vdots & k_2 \\ 0 & 0 & 1 & \vdots & k_3 \end{bmatrix}$	$\begin{bmatrix} 1 & 0 & 0 & \vdots & x \\ 0 & 1 & 0 & \vdots & y \\ 0 & 0 & 1 & \vdots & z \end{bmatrix}$

The matrix of coefficients is partially transformed to the identity matrix in the row echelon form and fully transformed in the reduced row echelon form.

Each step of the Gaussian elimination process uses one of three elementary row operations for matrices:

1. Swapping two rows.

2. Multiplying one row by a nonzero constant.

3. Multiplying one row by a nonzero constant and adding it to another row.

Not surprisingly, the HP 48 has a command corresponding to each of these operations. Take a look at an extended example of using Gaussian elimination and the HP 48 to solve a linear system.

---

*Technically, the method of reducing the augmented matrix to a row echelon matrix is called *Gaussian elimination* and the method that reduces the augmented matrix "all the way" to a reduced row echelon matrix is called *Gauss-Jordan reduction*. Both methods are referred to interchangeably as Gaussian elimination in this book.

Example: Solve this system of linear equations using Gaussian elimination:

$$x + 2y + 3z = 6$$
$$2x - 4y + 2z = 16$$
$$3x + y - z = -2$$

1. Enter the *augmented* matrix onto the stack. The augmented matrix includes an extra column containing the constants that appear on the

   right-hand side of the equations: $\begin{bmatrix} 1 & 2 & 3 & \vdots & 6 \\ 2 & -4 & 2 & \vdots & 16 \\ 3 & 1 & -1 & \vdots & -2 \end{bmatrix}$

   So, press →(MATRIX)(1)(ENTER)(2)(ENTER)(3)(ENTER)(6)(ENTER)(▼)(2)(ENTER)(4)(+/−)(ENTER)(2)(ENTER)(16)(ENTER)(3)(ENTER)(1)(ENTER)(1)(+/−)(ENTER)(2)(+/−)(ENTER)(ENTER).

2. The upper left element of the matrix does not need to be transformed, so begin with element (2,1)—the first element of the second row. It must be transformed to 0. To do this multiply row 1 by -2 and add it row 2: (2)(+/−)(ENTER)(1)(ENTER)(2)(ENTER)(MTH) **MATR ROW RCIJ**. Result:
   ```
   [[ 1  2  3  6 ]
    [ 0 -8 -4  4 ]
    [ 3  1 -1 -2 ]]
   ```

3. Reduce element (3,1) to 0: Multiply row 1 by -3; add it to row 3. (3)(+/−)(ENTER)(1)(ENTER)(3) **RCIJ**.

   Result:
   ```
   [[ 1  2  3  6 ]
    [ 0 -8 -4  4 ]
    [ 0 -5 -10 -20 ]]
   ```

4. Reduce element (2,2) to 1: Swap rows 2 and 3, then multiply row 2 by −1/5 (to avoid fractional elements). (2)(ENTER)(3)(NXT) **RSWP** (.)(2)(+/−)(ENTER)(2)(NXT) **RCI**. Result:
   ```
   [[ 1  2  3  6 ]
    [ 0  1  2  4 ]
    [ 0 -8 -4  4 ]]
   ```

5. Reduce element (2,3) to 0: Multiply row 2 by 8 and add it to row 3. (8)(ENTER)(2)(ENTER)(3) **RCIJ**. Result:
   ```
   [[ 1  2  3  6 ]
    [ 0  1  2  4 ]
    [ 0  0 12 36 ]]
   ```

6. Reduce element (3,3) to 1: Multiply row 3 by 1/12. $\boxed{1}$ $\boxed{\text{ENTER}}$ $\boxed{1}$ $\boxed{2}$ $\boxed{\div}$ $\boxed{3}$ ■RCI■. Result:
$$\begin{bmatrix} [1 & 2 & 3 & 6] \\ [0 & 1 & 2 & 4] \\ [0 & 0 & 1 & 3] \end{bmatrix}.$$

This produces the row echelon form, which, when translated it back

into a set of equations is
$$\begin{aligned} x + 2y + 3z &= 6 \\ y + 2z &= 4 \\ z &= 3 \end{aligned}$$

7. Of course, the above system could be now solved by substituting z = 3 into the second equation, then solving there for y, then substituting for y and z in the first equation, and solving there for x. But for the purposes of this example, continue the elimination process until you produce the completely reduced row echelon form.

Reduce element (1,2) to 0: Multiply row 2 by -2; add it to row 1. $\boxed{2}$ $\boxed{+/-}$ $\boxed{\text{ENTER}}$ $\boxed{2}$ $\boxed{\text{ENTER}}$ $\boxed{1}$ ■RCIJ■.

Result:
$$\begin{bmatrix} [1 & 0 & -1 & -2] \\ [0 & 1 & 2 & 4] \\ [0 & 0 & 1 & 3] \end{bmatrix}$$

8. Reduce element (2,3) to 0: Multiply row 3 by -2; add it to row 2. $\boxed{2}$ $\boxed{+/-}$ $\boxed{\text{ENTER}}$ $\boxed{3}$ $\boxed{\text{ENTER}}$ $\boxed{2}$ ■RCIJ■.

Result:
$$\begin{bmatrix} [1 & 0 & -1 & -2] \\ [0 & 1 & 0 & -2] \\ [0 & 0 & 1 & 3] \end{bmatrix}$$

9. Finally, reduce element (1,3) to 0: Multiply row 3 by 1 and add it to row 1: $\boxed{1}$ $\boxed{\text{ENTER}}$ $\boxed{3}$ $\boxed{\text{ENTER}}$ $\boxed{1}$ ■RCIJ■.

Result:
$$\begin{bmatrix} [1 & 0 & 0 & 1] \\ [0 & 1 & 0 & -2] \\ [0 & 0 & 1 & 3] \end{bmatrix}$$

The solution is now directly obvious if you translate this reduced row echelon form into a system of equations:

$$\begin{aligned} x &= 1 \\ y &= -2 \\ z &= 3 \end{aligned}$$

As the previous example makes clear, Gaussian elimination will get you to a solution sooner or later, but it may take more than a few steps. The HP 48 has a command to accelerate the process. RREF takes you directly from the augmented matrix to the reduced row echelon form; in essence, it automatically executes all of the elementary row operations necessary to complete the process.

Example: Use the RREF command to solve the following system:

$$x + 2y + 3z = 6$$
$$2x - 4y + 2z = 16$$
$$3x + y - z = -2$$

1. Enter the augmented matrix representing the system: →MATRIX 1
   ENTER 2 ENTER 3 ENTER 6 ENTER ▼ 2 ENTER 4 +/− ENTER 2
   ENTER 1 6 ENTER 3 ENTER 1 ENTER 1 +/− ENTER 2 +/− ENTER
   ENTER.

2. Transform the augmented matrix directly to its reduced row echelon form: MTH MATR FACTR RREF.

   Result:
   ```
   [[ 1 0 0  1 ]
    [ 0 1 0 -2 ]
    [ 0 0 1  3 ]]
   ```

3. As in the previous example, you can simply read the solution from the right-most column: $x = 1$; $y = -2$, $z = 3$.

The Gaussian elimination process can be used on any augmented matrix—even one that has been augmented with more than one column.

For example, to invert a square matrix, you can augment it with an identity matrix of the same dimensions and then reduce the augmented matrix to its reduced row echelon form. The inverse is returned to the right-hand section, just as the solution is returned to the right-hand section in the examples above.

Example: Use Gaussian elimination to invert $\begin{bmatrix} 1 & 2 & 3 \\ 1 & 1 & 2 \\ 0 & 1 & 2 \end{bmatrix}$

1. Enter the matrix: →(MATRIX)(1)(ENTER)(2)(ENTER)(3)(ENTER)(▼)(1)(ENTER)(1)(ENTER)(2)(ENTER)(0)(ENTER)(1)(ENTER)(2)(ENTER)(ENTER).

2. Create a 3 x 3 identity matrix: (3)(MTH) MATR MAKE IDN .

3. Augment the original matrix by inserting the identity matrix on the right side (i.e. beginning in column 4) of the original matrix: (4)(MTH) MATR COL COL+ . Result: [[1 2 3 1 0 0]
   [1 1 2 0 1 0]
   [0 1 2 0 0 1]]

4. Transform the augmented matrix to its reduced row echelon form: MATR FACTR RREF . Result: [[1 0 0 0 1 -1]
   [0 1 0 2 -2 -1]
   [0 0 1 -1 1 1]]

   Notice that the left half of the augmented matrix has been converted to a 3 x 3 identity matrix and the right half is the inverse of the original matrix.

5. Eliminate the first three columns, leaving only the inverse: (1)(MTH) MATR COL COL– (←)(1) COL– (←)(1) COL– (←).
   Result: [[0 1 -1]
   [2 -2 -1]
   [-1 1 1]]

Solving with Determinants: Cramer's Rule

Determinants can be used to solve a system of linear equations provided that the following three conditions are met:

- The number of independent variables equals the number of independent equations in the system.
- The determinant of the matrix of coefficients is non-zero.
- At least one of the constants to the right of the equal signs is non-zero.

Cramer's Rule requires that you create a set of specially augmented matrices from the matrix of coefficients. For example, to use Cramer's Rule to solve the linear system $\begin{array}{l} x+2y+3z = 6 \\ 2x-4y+2z = 16 \\ 3x+y-z = -2 \end{array}$, you first must create *four* matrices:

- The matrix of coefficients itself: $\mathbf{A} = \begin{bmatrix} 1 & 2 & 3 \\ 2 & -4 & 2 \\ 3 & 1 & -1 \end{bmatrix}$

- A with its first column replaced by the constants: $\mathbf{A}_x = \begin{bmatrix} 6 & 2 & 3 \\ 16 & -4 & 2 \\ -2 & 1 & -1 \end{bmatrix}$

- A with its second column replaced by the constants: $\mathbf{A}_y = \begin{bmatrix} 1 & 6 & 3 \\ 2 & 16 & 2 \\ 3 & -2 & -1 \end{bmatrix}$

- A with its third column replaced by the constants: $\mathbf{A}_z = \begin{bmatrix} 1 & 2 & 6 \\ 2 & -4 & 16 \\ 3 & 1 & -2 \end{bmatrix}$

The unique solution of the given system is: $x = \dfrac{|\mathbf{A}_x|}{|\mathbf{A}|} \qquad y = \dfrac{|\mathbf{A}_y|}{|\mathbf{A}|} \qquad z = \dfrac{|\mathbf{A}_z|}{|\mathbf{A}|}$,

where | | indicates the *determinant* of the respective matrix.

Example: Solve the following system of linear equations using Cramer's Rule:

$$x + 2y - z = -7$$
$$2x + 3y + 2z = -3$$
$$x - 2y - 2z = 3$$

1. Enter the augmented matrix, just as with Gaussian elimination: →
   MATRIX 1 ENTER 2 ENTER 1 +/− ENTER 7 +/− ENTER ▼ 2
   ENTER 3 ENTER 2 ENTER 3 +/− ENTER 1 ENTER 2 +/− ENTER
   2 +/− ENTER 3 ENTER ENTER.

2. Copy the augmented matrix, then modify the copy by removing the
   column of constants: ENTER 4 MTH **MATR COL COL−** ←.
   Result: [[1 2 −1]
                 [2 3 2]   The matrix of coefficients (**A**)
                 [1 −2 −2]]

3. Find the determinant of A: **MATR NORM** NXT **DET**. Result: 17

4. Bring the copy of the augmented matrix to level 1 and make another
   copy (SWAP ENTER). Then create A_x by swapping column 4 and
   column 1 and then deleting column 4: 1 ENTER 4 **MATR COL**
   CSWP 4 **COL−** ←. Result: [[−7 2 −1]
                                          [−3 3 2]
                                          [3 −2 −2]]

5. The determinant of A_x: **MATR NORM** NXT **DET**. Result: 17

6. Repeat steps 4 and 5 to find the determinant of A_y. This time swap
   columns 2 and 4: SWAP ENTER 2 ENTER 4 **MATR COL CSWP**
   4 **COL−** ← **MATR NORM** NXT **DET**. Result: −51

7. Likewise, find the determinant of A_z. This time swap columns 3 and
   4: SWAP 3 ENTER 4 **MATR COL CSWP** 4 **COL−** ← **MATR**
   NORM NXT **DET**. Result: 34

8. Collect the last three determinants in a list: 3 PRG **LIST →LIST**.

   Result: 2:   17
                1:          { 17 −51 34 }

9. To compute the three solutions, swap levels and divide: SWAP ÷.
   Result: { 1 −3 2 }   Thus, $x = 1$; $y = -3$; $z = 2$.

Of course, this method of solving a system of linear equations by Cramer's Rule is a good candidate for automation. The program CRAMER (see page 279) does just that.

Example: Use CRAMER to solve the following linear system:

$$
\begin{aligned}
a + b + c + d + f &= 340 \\
a + b &= 90 \\
a + \quad c &= 110 \\
d + f &= 180 \\
c + \quad f &= 170
\end{aligned}
$$

1. Enter the augmented matrix representing the system: [→]MATRIX[1] [ENTER][1][ENTER][1][ENTER][1][ENTER][1][ENTER][3][4][0][ENTER][▼][1] [ENTER][1][ENTER][0][ENTER][0][ENTER][0][ENTER][9][0][ENTER] [1][ENTER][0][ENTER][1][ENTER][0][ENTER][0][ENTER][1][1][0][ENTER] [0][ENTER][0][ENTER][0][ENTER][1][ENTER][1][ENTER][1][8][0][ENTER] [0][ENTER][0][ENTER][1][ENTER][0][ENTER][1][ENTER][1][7][0][ENTER] [ENTER].

2. Enter a list of the variables, in the order presented in the matrix: [←] [{ }][α][←][A][SPC][α][←][B][SPC][α][←][C][SPC][α][←][D][SPC][α][←][F] [ENTER].

3. Execute CRAMER: [α][α][C][R][A][M][E][R][ENTER] or [VAR] (then [NXT] or [←][PREV] as needed) CRAM.

   Result:
   ```
   2:  { -1 -40 -50 -70 -80 -100 }
   1:  { :a: 40 :b: 50 :c: 70 :d: 80 :f:
        100 }.
   ```

The list on level 2 contains the determinants for the matrix of coefficients and each of the "Cramer"-augmented matrices. Level 1 contains a list of the solutions tagged with the names of the variables.

Solving by Matrix Inversion

While the preceding methods are robust (and could be adapted to symbolic—as opposed to numeric—solutions) they are not the most efficient means of solving a system of linear equations when using a computational device like the HP 48. The fastest methods usually employ algorithms to invert a matrix.

The crucial role that matrix inversion plays in the solution of a system of linear equations is obvious when you view a linear system as a *single* matrix equation:

$$\begin{array}{c} x+2y-z=-7 \\ 2x+3y+2z=-3 \\ x-2y-2z=3 \end{array} \quad \text{is equivalent to} \quad \begin{bmatrix} 1 & 2 & -1 \\ 2 & 3 & 2 \\ 1 & -2 & -2 \end{bmatrix} \cdot \begin{bmatrix} x \\ y \\ z \end{bmatrix} = \begin{bmatrix} -7 \\ -3 \\ 3 \end{bmatrix}$$

If you recall the rules of matrix multiplication, you will see that this relationship is exactly correct. The matrix equation can be simplified to $\mathbf{A} \cdot \mathbf{x} = \mathbf{B}$, where $\mathbf{A}$ is the matrix of coefficients, $\mathbf{x}$ is the vector of variables and $\mathbf{B}$ is the matrix of constants.

To solve the matrix equation, you must multiply both sides of the equation by the *inverse* of $\mathbf{A}$, as follows: $\mathbf{A}^{-1} \cdot \mathbf{A} \cdot \mathbf{x} = \mathbf{A}^{-1} \cdot \mathbf{B} = \mathbf{I} \cdot \mathbf{x} = \mathbf{x} = \mathbf{A}^{-1} \cdot \mathbf{B}$. Thus the solution can be computed by multiplying the inverse of $\mathbf{A}$ by the vector of constants.

You've already seen two methods of computing the inverse: using the $\boxed{1/x}$ key and using Gaussian elimination (the routine used by the HP 48 when you press $\boxed{1/x}$ to inverse a matrix makes use of advanced matrix decomposition algorithms that are beyond the scope of this book to explain). But there is a third method for inverting a matrix that you should know about: the *cofactor matrix*. The cofactor matrix of a given square matrix is the square matrix in which each element is replaced by its respective *cofactor*.

So... what's a cofactor?

To put it as simply as possible: Each element in a matrix belongs to a particular row i and a particular column j. That element's *cofactor* is $(-1)^{i+j}$ times the determinant of the matrix that remains if you remove row i and column j.

For example, given $\mathbf{A} = \begin{bmatrix} 1 & 2 & -1 \\ 2 & 3 & 2 \\ 1 & -2 & -2 \end{bmatrix}$, the cofactor of the element in row 2,

column 3, is $(-1)^{2+3}\begin{vmatrix} 1 & 2 \\ 1 & -2 \end{vmatrix} = 4$. And likewise, if you find the cofactors for every

element in $\mathbf{A}$, you'll have its complete cofactor matrix: $\mathbf{A}^c = \begin{bmatrix} -2 & 6 & -7 \\ 6 & -1 & 4 \\ 7 & -4 & -1 \end{bmatrix}$

As it turns out, the inverse of a matrix $\mathbf{A}$ is the *transpose* of its cofactor matrix, divided by the determinant of $\mathbf{A}$. Thus, for the example matrix:

$$\mathbf{A}^{-1} = \frac{\{\mathbf{A}^c\}^T}{|\mathbf{A}|} = \frac{\begin{bmatrix} -2 & 6 & 7 \\ 6 & -1 & -4 \\ -7 & 4 & -1 \end{bmatrix}}{17} = \begin{bmatrix} -\dfrac{2}{17} & \dfrac{6}{17} & \dfrac{7}{17} \\ \dfrac{6}{17} & -\dfrac{1}{17} & -\dfrac{4}{17} \\ -\dfrac{7}{17} & \dfrac{4}{17} & -\dfrac{1}{17} \end{bmatrix}$$

Compare this to the result when you use $\boxed{1/x}$ to compute the inverse and use $\mathbf{A} \rightarrow \mathbf{Q}$ to rationalize the elements:

```
{ { '-(2/17)'  '6/17'  '7/17' }
  { '6/17'  '-(1/17)'  '-(4/17) }
  { '-(7/17)  '4/17'  '-(1/17) } }
```

The program COFACTR (see page 278) computes the cofactor matrix for a given square matrix. Try an example using it to solve a linear system....

Example: Use COFACTR to help solve

$$x + 2y + z - 6t = 12$$
$$2x + 3y - 2z + t = 10$$
$$-3x - 4y + 3z + 5t = 3$$
$$4x - 3y - z + t = 6$$

1. The augmented matrix: →[MATRIX] 1 [SPC] 2 [SPC] 1 [SPC] 6 +/−
[SPC] 1 2 [ENTER] ▼ 2 [SPC] 3 [SPC] 2 +/− [SPC] 1 [SPC] 1 0 [ENTER]
3 +/− [SPC] 4 +/− [SPC] 3 [SPC] 5 [SPC] 3 [ENTER] 4 [SPC] 3 +/−
[SPC] 1 +/− [SPC] 1 [SPC] 6 [ENTER] [ENTER].

2. Extract the column of constants from the matrix (column 5): 5 [MTH]
MATR COL COL−.

3. The column of constants will be used later; swap it to level 2: [SWAP].
Then with the matrix of coefficients on level 1, make a copy and execute the COFACTR program to create the cofactor matrix: [ENTER] α
α C O F A C T R [ENTER] or [VAR] (NXT) or ←[PREV] as needed)
COFAC. Result: [[59 39 116 −3]
                         [73 85 93 56]
                         [46 35 118 39]
                         [51 −26 13 2]]

4. Transpose the cofactor matrix: [MTH] **MATR MAKE TRN**.

5. Grab the matrix of coefficients now sitting on level 2 and find its
determinant: [SWAP] [MTH] **MATR NORM** (NXT) **DET**. Result: 271

6. Divide the transposed cofactor matrix by the determinant, make a
copy and rationalize the result: ÷ [ENTER] [VAR] **A→Q**. Result:

{ { '59/271' '73/271' '46/271' '51/271' }
  { '39/271' '85/271' '35/271' '−(26/271)' }
  { '116/271' '93/271' '118/271' '13/271' }
  { '−(3/271)' '56/271' '39/271' '2/271' } }

This is the inverse matrix of **A**.

7. Drop the rationalized version, then multiply the inverse matrix by
the matrix of constants extracted in step 2. You must swap first to put
the matrices in the proper order ($A^{-1} \cdot B = x$): ← [SWAP] ×.
Result: [6.9446 4.6753 10.1624 2.4096]

So, $x = 6.9446$; $y = 4.6753$; $z = 10.1624$; and $t = 2.4096$.

Finally, here are some examples to illustrate the built-in routine for solving linear systems, which uses matrix inversion to solve the matrix equation: $\mathbf{A} \cdot \mathbf{x} = \mathbf{B}$.

Example: Solve this linear system using the stack:
$$x + 2y + z - 6t = 12$$
$$2x + 3y - 2z + t = 10$$
$$-3x - 4y + 3z + 5t = 3$$
$$4x - 3y - z + t = 6$$

1. Enter the array of constants (**B**): ⟵[ ] 1 2 SPC 1 0 SPC 3 SPC 6 ENTER.

2. Enter the matrix of coefficients (**A**): →MATRIX 1 ENTER 2 ENTER 1 ENTER 6 +/− ENTER ▼ 2 ENTER 3 ENTER 2 +/− ENTER 1 ENTER 3 +/− ENTER 4 +/− ENTER 3 ENTER 5 ENTER 4 ENTER 3 +/− ENTER 1 +/− ENTER 1 ENTER ENTER.

3. Divide the two arrays: ÷ This performs the INVerse operation (as in 1/x) on the level 1 matrix and multiplies the result by the level 2 array. <u>Result</u>: **[6.9446 4.6753 10.1624 2.4096]**

Example: Use the **SOLVE SYSTEM** application to compute the solution to the same system as in the previous example.

1. Open the **SOLVE SYSTEM** application: →SOLVE ▲ ▲ ENTER.

2. Enter the coefficients matrix in the **A:** field (same as step 2 above).

3. Enter the constants array in the **B:** field (same as step 1 above).

4. Highlight the **X:** field, and **SOLVE**.

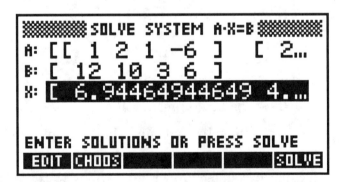

5. Press **EDIT**, then ▶ as needed to clarify the results. Use CANCEL CANCEL to see the stack and the result array, labeled **Solutions:**.

Symbolic Solutions of Linear Systems

All of the built-in matrix operations, solving commands, and programs discussed so far in this chapter require that you use numeric arrays; algebraic objects are not allowed in such solutions on the HP 48.

However, you can extend the principles of the matrix operations and commands to "symbolic" matrices by devising a method to represent symbolic matrices on the HP 48. The generally accepted method for doing this is the "list-of-lists" notation.

For example, the linear system

$$ax + by + c = d$$
$$ex + fy + gz = h$$
$$jx + ky + lz = m$$

can be represented (in augmented form) as:

```
{{ a b c d }
 { e f g h }
 { j k l m }}
```

Using the same underlying mathematical principles as the numeric procedures, a set of programs has been designed to work with the "list of lists" form of symbolic matrices, performing essential matrix operations discussed in this chapter—matrix arithmetic, transposition, inversion, finding determinants, and solving systems of linear equations.* Like the other programs referred to in this book, these programs are listed in alphabetically order in the appendix at the end of the book.

*Some of the programs in this set—SDET, SMMULT, STRN, SM→, →SM, SCOF, SMADD, SMSMULT, and SMSUB— were adapted from the set of programs developed by Bill Wickes in his book, *HP 48 Insights, Part I: Principles and Programming* and are used here with his permission.

Here is a summary of the symbolic matrix programs* and how they work (along with the page numbers where they are listed in the Appendix):

Symbolic Matrix Arithmetic

SMADD   Sums two symbolic matrices of equal dimensions (page 305).

SMSUB   Subtracts one symbolic matrix from another (page 307).

SMSMULT   Multiplies a scalar (which may also be symbolic) by a symbolic matrix (page 306).

SMMULT   Multiples two symbolic matrices provided that they are dimensionally compatible (page 305).

Symbolic Matrix Operations

SM→   Disassembles a symbolic matrix onto the stack, with each element occupying its own stack level. The number of columns in the matrix is returned to level 2, and the number of rows is returned to level 1. This program is analogous to the built-in OBJ→ command for numeric arrays (page 304).

→SM   Assembles a symbolic matrix from its elements on the stack. Level 2 should have the number of columns in the new matrix and level 1 should have the number of rows. This is analogous to the built-in →ARRY command for numeric arrays (page 305).

STRN   Transposes a symbolic matrix (page 313).

SDET   Finds the determinant of a square symbolic matrix (page 302).

SCOF   Finds the cofactor for an element in row r (level 2), column c (level 1) of a square symbolic matrix (level 3). (page 301)

SRSWP   Swaps two rows of a symbolic matrix (page 312).

SRCI   Multiplies a row of a symbolic matrix by a constant (which may be symbolic). Analogous to RCI for numeric matrices (page 312).

*Note that you may enter a numeric matrix as a symbolic matrix if you want to perform operations in combination with a truly symbolic matrix.

5. SYSTEMS OF LINEAR EQUATIONS

SRIJ	Multiplies a row of a symbolic matrix by a constant (which may also be symbolic) and adds the product to a second row. Analogous to RCIJ for numeric matrices (page 312).
SXROW	Extracts a designated row from a symbolic matrix, leaving it on leaving it on level and the diminished matrix on level 2. Analogous to ROW- for numeric matrices (page 313).
SNROW	Inserts a row list or symbolic matrix into a symbolic matrix beginning in the designated position. Analogous to ROW+ for numeric matrices (page 307).
SCSWP	Swaps two columns of a symbolic matrix (page 302).
SXCOL	Extracts a designated column from a symbolic matrix, leaving it as a *row* list on level 1, and the diminished matrix on level 2. Analogous to COL- for numeric matrices (page 313).
SNCOL	Inserts a *row* list representing a column of elements or a symbolic matrix into a symbolic matrix beginning in the designated position. Analogous to COL+ for numeric matrices (page 307).

Symbolic Linear Solutions

SCRAMER	Given an augmented symbolic matrix representing a linear system, returns a list of solutions computed using Cramer's Rule. Analogous to CRAMER for numeric matrices Uses SDET, SCOF, and SMSMULT (page 302).
SCFACTR	Computes the symbolic cofactor matrix for a square symbolic matrix. Analogous to COFACTR for numeric matrices. Uses SDET and SCOF (page 301).
SMINV	Finds the inverse of a square symbolic matrix, using the cofactor matrix algorithm. Uses SCFACTR (page 305).
SMSOLV	Solves a linear system represented by an augmented symbolic matrix, using the cofactor matrix algorithm. Uses SMINV and SMMULT (page 306).

Over- and Under-Determined Systems

All of the examples and techniques shown so far in the chapter have used *exactly-determined* systems of linear equations—systems with equal numbers of independent equations and independent variables. But you may run across systems where this is not the case.

The HP 48 provides a special command for handling situations, where attempts fail to recast the linear system as exactly determined. The command LSQ finds the *closest* solution—the one which results in the smallest least-squares error. Or, if LSQ finds more than one equivalent solution, it returns that with the smallest Euclidean norm (the array's "absolute value"). Look at two cases—one over-determined system and one under-determined system (these examples apply only to numeric matrices; there is no symbolic equivalent provided in this book):

Example: Find the *best* solution to this system:

$$x + 2y - 3z = 34$$
$$-3x + y + 5z = 21$$
$$4x - y + 2z = 20$$
$$-x - 4y + 7z = 15$$

1. Enter the vector of constants: ←[[ ]] [3][4] [SPC] [2][1] [SPC] [2][0] [SPC] [1][5] [ENTER].

2. Enter the matrix of coefficients: [→][MATRIX] [1] [ENTER] [2] [ENTER] [3] [+/−] [ENTER] [▼] [3][+/−] [ENTER] [1] [ENTER] [5] [ENTER] [4] [ENTER] [1][+/−] [ENTER] [2] [ENTER] [1][+/−] [ENTER] [4][+/−] [ENTER] [7] [ENTER] [ENTER].

3. Solve the over-determined system: [MTH] **MATR** **LSQ**.

   Result (to 4 places): **[5.4725 9.1617 6.0350]**

Example: Find the best solution for this linear system:

$$x + 2y - 3z = 34$$
$$-3x + y + 5z = 21$$

1. Enter the vector of constants: ←[[ ]] [3][4] [SPC] [2][1] [ENTER].

2. Enter the matrix of coefficients: [→][MATRIX] [1] [ENTER] [2] [ENTER] [3][+/−] [ENTER] [▼] [3][+/−] [ENTER] [1] [ENTER] [5] [ENTER] [ENTER].

3. Solve the under-determined system: [MTH] **MATR** **LSQ**.

   Result: **[−4.2222 16.6239 −1.6581]**

Systems of Linear Inequalities

Systems of linear inequalities differ from systems of linear equations primarily in the nature of their solution sets. Exactly-determined systems result in a unique solution—in essence, a *point*. But a system of inequalities results in a solution *space*—an infinite number of points, where every point satisfies *all of the inequalities in the system*. Note that such systems are not said to be either over- or under-determined; the ratio of equations to variables is unimportant.

For systems in two variables, plotting is a good means of determining the solution space. The program, INPLOT (see page 284), by Jim Donnelly, does this.*

Example: Plot this set of linear inequalities:
$$Y \geq 2X - 1$$
$$Y \leq -2X - 2$$
$$Y \leq 3X + 2$$
$$Y \geq -2$$

1. Open the **PLOT** application; set the **TYPE:** field to Function.

2. Highlight the **EQ:** field; enter the inequalities in a list: ⦅←⦆⦅{}⦆⦅'⦆⦅α⦆⦅Y⦆ ⦅α⦆⦅→⦆⦅3⦆⦅2⦆⦅X⦆⦅α⦆⦅X⦆⦅−⦆⦅1⦆⦅▶⦆⦅'⦆⦅α⦆⦅Y⦆⦅α⦆⦅←⦆⦅3⦆⦅2⦆⦅+/−⦆⦅X⦆⦅α⦆⦅X⦆⦅−⦆⦅2⦆⦅▶⦆ ⦅'⦆⦅α⦆⦅Y⦆⦅α⦆⦅←⦆⦅3⦆⦅3⦆⦅X⦆⦅α⦆⦅X⦆⦅+⦆⦅2⦆⦅▶⦆⦅'⦆⦅α⦆⦅Y⦆⦅α⦆⦅→⦆⦅3⦆⦅2⦆⦅+/−⦆⦅ENTER⦆.

3. Set the plot parameters—**INDEP:** to X, **H-VIEW** to −2 2 and **V-VIEW** to −3 3. Save the settings and exit to the stack: ⦅NXT⦆ ▮OK▮.

4. Plot the system using the INPLOT program: ⦅α⦆⦅α⦆⦅I⦆⦅N⦆⦅P⦆⦅L⦆⦅O⦆⦅T⦆ ⦅ENTER⦆ or ⦅VAR⦆ (then ⦅NXT⦆ or ⦅←⦆⦅PREV⦆ as needed) ▮INPLO▮.**

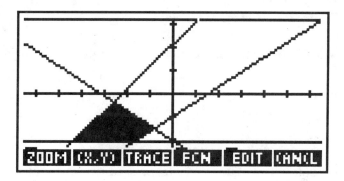

*Note that INPLOT requires that you use X and Y (uppercase) as the independent and dependent variables.
**Plotting a system of inequalities is relatively slow because each column of pixels must be tested against the values of the each of the functions. The line of pixels along the top shows you the progress of the plotting.

Linear Programming

How do you solve a system of linear inequalities involving more than two independent variables, when it isn't possible to plot the solution?

To be sure, there is no easy approach. However, most of the real-world applications for solving such systems of linear inequalities are found in the context of finding the *optimum* solution within a possible solution space. In these cases, you are not interested in the entire solution space (or even in defining it), but in determining the *one solution* within the universe of possible solutions that optimizes a particular function. Such problems are the realm of *linear programming*.

A linear programming problem consists of:

- An *objective function* that must be optimized—maximized or minimized.

- A set of *constraints*—linear inequalities that define the possible solution space for the problem.

Solving a linear programming problem requires that you find a means to "graph" the set of constraints. If there are only two independent variables in the system, the "graph" is a two-dimensional polygon and can be actually drawn on a piece of paper or on the HP 48 screen. If there are three independent variables, the "graph" is a three-dimensional polyhedron and might be sculpted as a model or be represented in 3D-rendering on a flat surface. However, the "graph" of a system containing four or more variables cannot be created in any physical way within in the three dimensions of our existence. So how can you solve a linear programming problem with four or more variables?

Problems in four or more dimensions can't be represented physically, only *mathematically,* via matrices. If a linear programming problem can be represented in terms of matrices, then there is a good chance that it can be solved—even if there are 5, 10, or morevariables.

What does the "graph" of the solution space of a system with, say, 7 independent variables "look" like? The solution space of a 3-variable solution is a 3-dimensional polyhedron, but the mathematical term for a higher-dimensional analogue of a polyhedron is *simplex*. Thus, the solution space of a 7-variable system is a 7-dimensional simplex.

Recall that to graph any inequality, you must actually graph the corresponding equality (a line) and then decide on which "side" of the line the solution space lies. Similarly, the "edges" of any solution space for a system of linear inequalities— be it a polygon, polyhedron, or simplex—are the inequalities of the linear system *after* they are converted into equalities. These solution spaces all have *vertices*, which are the intersection points of two or more "edges" (i.e. inequalities converted into equations).

Now, one of central theorems of linear programming is that the optimal solution, if it exists, will always occurs at a vertex of the solution space that represents the set of constraints for the problem. Therefore, the process of "solving" a linear programming problem is simply the testing of the vertices of a simplex to see which of them—when its coordinates are substituted into the objective function —yields the optimum value (maximum or minimum, depending on the problem).

However, to solve a linear programming problem before the end of the millennium, you must test the vertices in an efficient way; it takes far too long to test every vertex, so you need to test only those vertices that might yield the optimum and ignore those that don't stand a chance. (And obviously, this need for efficiency increases rapidly as the number of variables—the number of dimensions in the simplex—increases).

The *Simplex Method* is a matrix-based algorithm that explores the vertices of the solution simplex in an very efficient way. It starts with any vertex and then finds another vertex that improves the objective function, and repeats the procedure until there are no more improvements. It almost always finds the optimal vertex after just a few iterations—no more than the number of inequalities or the number of independent variables, whichever is larger.

The HP 48 can be programmed to use the Simplex method. The examples which follow use a collection of four programs:

- LINPRG (see page 286) is the master controlling program, collecting a description of the linear programming problem, converting it to the array form—the *tableau*—needed by the 2-phase Simplex algorithm, calling PHASE1 and SIMPLEX as needed to solve the problem, and finally reporting the results.

- PHASE1 (see page 292) adjusts the given tableau so that it is in canonical maximum form, suitable for the main SIMPLEX algorithm to search for an optimum solution.

- SIMPLEX (see page 303) applies the Simplex algorithm to the given tableau, using a specified objective function.

- PIVOT (see page 294), called by both PHASE1 and SIMPLEX, performs a single pivot operation on the given tableau using a specified *pivot row* and *pivot column*.

Although PHASE1, SIMPLEX, and PIVOT are primarily designed for use with LINPRG, you may also use them independently if you want to examine the process more closely.*

Look at some examples using LINPRG. Be sure that all four programs are correctly entered in your current path before trying these examples.

---

*SIMPLEX takes as inputs: the number of constraints (level 6), the number of decision variables (level 5), a list of the indexes for the variables in the current solution—*basis variables* (level 4), a list of the indexes for the variables not in the current solution—*non-basis variables* (level 3), an array representing the current Tucker tableau (level 2), and either the value 2, if the tableau is non-canonical, or 1 if canonical (level 1). Note that the order of the index lists on levels 2 and 3 must correspond to the ordering of elements in the Tucker tableau and that the index, 0, is used for any artificial variables in the tableau. PHASE1 takes the same inputs and in the same order as SIMPLEX with the exception of the final input (level 1 for SIMPLEX). PIVOT takes the same five inputs as PHASE1, in the same order, moving them to levels 7 through 3, and additionally takes the pivot row (level 2) and the pivot column (level 1). SIMPLEX and PIVOT return the number of constraints to level 5, the number of decision variables to level 4, a revised list of the indexes for the basis variables to level 3, a revised list of the indexes for the non-basis variables to level 2, and a revised Tucker tableau to level 1. PHASE1 does not return anything to the stack, but (as SIMPLEX and PIVOT do as well) returns the revised list of basis indexes to the variable bvars, the revised list of non-basis indexes to the variable nvars, and the revised Tucker tableau to the variable a.

Example: An investment manager wants to invest $20,000 each month for a client in the bond market. He has three kinds of bonds (i.e. three different risk categories) that he may consider. The return on each kind varies from month to month, but this month he can get 7% on the safest kind of bond, 8.5% on the riskiest kind of bond he's allowed to consider, and 8% on the moderately risky kind. He need not invest the entire fund, but he can invest no more than 40% of the amount invested in any one type of bond. Further, he must invest at least $4000 in the safest kind of bond. To maximize the return on his investment, how should he allocate the investment this month?

1. *Define the variables.* The variables here are the amount invested in each type of bond. Thus, there are three variables: b_1, b_2, and b_3.

2. *Find the objective function.* This is the function that computes the return on the investment: $0.07b_1 + 0.08b_2 + 0.085b_3$

3. *Find the set of constraints.* In most real-world LP problems, the most common constraint is that all variables must be nonnegative ($b_1 \geq 0$, $b_2 \geq 0$, and $b_3 \geq 0$). And in fact, the SIMPLEX program *assumes* that all variables are nonnegative. The other constraints here are:

$$b_1 + b_2 + b_3 \leq 20,000$$
$$b_1 \leq 0.4(b_1 + b_2 + b_3)$$
$$b_2 \leq 0.4(b_1 + b_2 + b_3)$$
$$b_3 \leq 0.4(b_1 + b_2 + b_3)$$
$$b_1 \geq 4000$$

However, before you can use the LINPRG program, you must express each of the constraints so that all the variable terms are on the left side of the inequality and the constant is on the right side (the first and last constraints are already in proper form). Rearrange the constraints as necessary to the form needed by LINPRG:

$$b_1 + b_2 + b_3 \leq 20,000$$
$$0.6b_1 - 0.4b_2 - 0.4b_3 \leq 0$$
$$-0.4b_1 + 0.6b_2 - 0.4b_3 \leq 0$$
$$-0.4b_1 - 0.4b_2 + 0.6b_3 \leq 0$$
$$b_1 \geq 4000$$

4. Begin LINPRG by typing ⍺⍺LINPRG ENTER or pressing VAR (then NXT or ←PREV as needed) LINPR.

```
██████ LINEAR PROGRAMMING ██████
OBJECTIVE: ███████████████
CONSTRAINTS:
VARS:
MAX OR MIN?     "MAX"

ENTER OBJECTIVE FUNCTION
▐ EDIT ▌      ▐      ▌  ▐CANCL▌▐ OK ▌
```

5. Enter the objective function into the OBJECTIVE field:

$$0.07b_1 + 0.08b_2 + 0.085b_3 = 0$$

Press ' • 0 7 × ⍺ ← B 1 + • 0 8 × ⍺ ← B 2 + • 0 8 5 × ⍺ ← B 3 ← = 0 ENTER. Note that the objective function must have a right-hand constant, just like the constraints. If there is no constant in the expression, use zero.

6. Enter the set of constraints as a list: ← { } ' ⍺ ← B 1 + ⍺ ← B 2 + ⍺ ← B 3 ⍺ ← 3 2 0 0 0 0 ▶ ' • 6 × ⍺ ← B 1 — • 4 × ⍺ ← B 2 — • 4 × ⍺ ← B 3 ⍺ ← 3 0 ▶ ' • 4 +/− × ⍺ ← B 1 + • 6 × ⍺ ← B 2 — • 4 × ⍺ ← B 3 ⍺ ← 3 0 ▶ ' • 4 +/− × ⍺ ← B 1 — • 4 × ⍺ ← B 2 + • 6 × ⍺ ← B 3 ⍺ ← 3 0 ▶ ' ⍺ ← B 1 ⍺ → 3 4 0 0 0 ENTER.

7. Enter the list of variables, in the same order as they are in the constraints and in the objective function: ← { } ⍺ ← B 1 SPC ⍺ ← B 2 SPC ⍺ ← B 3 ENTER.

8. Since you do wish to maximize the objective function, you need not change the entry in the last field. Simply press ▐ OK ▌ to compute the solution. After a bit, you will see a message box:

Solution found. Press ▐ OK ▌ again.

Result: 3: { 5 3 2 6 1 }
2: [[0 1 .4 4000.000002]
    [1 0 .4 8000.000002]
    [-1 1 .6 7999.999998]
    [1 -1 -.2 .000004]
    [0 -1 0 4000]
    [-.005 -.01 -.082 -1599.99999563]
    [0 0 0 0]]
1: { :b1: 4000 :b2: 7999.999998 :b3:
    8000.000002 }

A list of tagged values for each of the decision variables—the solution—is returned to level 1. Note that small round-off errors will show up in computed values. Choosing to fix the display to an appropriate degree of precision for the solution you're determing (eg. to two places for problems dealing with money) is usually a good idea. Thus, the manager would invest $4000 in bond type 1, $8000 in bond type 2, and $8000 in bond type 3 in order to maximize the return during the month in question.

The objects on levels 2 and 3 are there to help you evaluate the quality of the solution, if you so choose, and can be dropped if you do not so choose. The final tableau is returned to level 2. The list on level 3 contains, in order, the indexes of the variables used in the solution—the *basis variables*. Because there are three decision variables in this problem—*b1*, *b2*, and *b3*—the smallest three non-zero indexes (1, 2, and 3) refer to them: All three figure in the solution in this case. Indexes higher than the number of decision variables or zero reflect *slack* and *artificial* variables created in the process of solving the problem and are not a part of the solution, even if they end up in the basis list.*

Try another example....

---

*Slack variables are useful sometimes in sensitivity analysis—the process of determining how sensitive the solution you computed is to small changes in the original constraints or objective function. However, this is beyond the scope of this book.

Example: A herd of livestock requires weekly at least 450 pounds of protein, 400 pounds of carbohydrates, and 1050 pounds of roughage. A bale of hay has 10 pounds of protein, 10 pounds of carbohydrates, 60 pounds of roughage, and costs $3.80. A sack of oats has 15 pounds of protein, 10 pounds of protein, 25 pounds of roughage, and costs $5.00. A sack of pellets has 10 pounds of protein, 5 pounds of carbohydrates, 55 pounds of roughage, and costs $3.50. A sack of sweet feed has 25 pounds of protein, 20 pounds of carbohydrate, 35 pounds of roughage, and costs $8.00.   How would you adequately feed the herd at minimum cost?

1. *Define the variables.* The variables here are the amounts of each food source: bales of hay (h), sacks of oats (o), sacks of pellets (p), and sacks of sweet feed (s).

2. *Find the objective function.* The sum of costs of the food:
$$3.80h + 5.00o + 3.50p + 8.00s$$

3. *Find the set of constraints.* The set of constraints incorporate the minimum weekly requirements for the three categories of nutrients: protein, carbohydrates, and roughage:
$$10h + 15o + 10p + 25s \geq 450$$
$$10h + 10o + 5p + 20s \geq 400$$
$$60h + 25o + 55p + 35s \geq 1050$$

4. *Enter the LP problem into the* LINPRG *and compute the solution, if one exists.* Begin LINPRG by typing $\boxed{\alpha}\boxed{\alpha}\boxed{L}\boxed{I}\boxed{N}\boxed{P}\boxed{R}\boxed{G}\boxed{\text{ENTER}}$ or pressing $\boxed{\text{VAR}}$ (then $\boxed{\text{NXT}}$ or $\boxed{\leftarrow}\boxed{\text{PREV}}$ as needed) **LINPR**.

5. Enter the objective function in the **OBJECTIVE** field, including the right-hand constant (0 here): $3.80h + 5.00o + 3.50p + 8.00s = 0$
Press $\boxed{\text{'}}\boxed{3}\boxed{.}\boxed{8}\boxed{\times}\boxed{\alpha}\boxed{\leftarrow}\boxed{H}\boxed{+}\boxed{5}\boxed{\times}\boxed{\alpha}\boxed{\leftarrow}\boxed{O}\boxed{+}\boxed{3}\boxed{.}\boxed{5}\boxed{\times}\boxed{\alpha}\boxed{\leftarrow}\boxed{P}$
$\boxed{+}\boxed{8}\boxed{\times}\boxed{\alpha}\boxed{\leftarrow}\boxed{S}\boxed{\leftarrow}\boxed{=}\boxed{0}\boxed{\text{ENTER}}$.

6. Enter the set of constraints (a list): $\boxed{\leftarrow}\boxed{\{\,\}}\boxed{\text{'}}\boxed{1}\boxed{0}\boxed{\times}\boxed{\alpha}\boxed{\leftarrow}\boxed{H}\boxed{+}\boxed{1}\boxed{5}$
$\boxed{\times}\boxed{\alpha}\boxed{\leftarrow}\boxed{O}\boxed{+}\boxed{1}\boxed{0}\boxed{\times}\boxed{\alpha}\boxed{\leftarrow}\boxed{P}\boxed{+}\boxed{2}\boxed{5}\boxed{\times}\boxed{\alpha}\boxed{\leftarrow}\boxed{S}\boxed{\alpha}\boxed{\rightarrow}\boxed{3}\boxed{4}\boxed{5}\boxed{0}$
$\boxed{\blacktriangleright}\boxed{\text{'}}\boxed{1}\boxed{0}\boxed{\times}\boxed{\alpha}\boxed{\leftarrow}\boxed{H}\boxed{+}\boxed{1}\boxed{0}\boxed{\times}\boxed{\alpha}\boxed{\leftarrow}\boxed{O}\boxed{+}\boxed{5}\boxed{\times}\boxed{\alpha}\boxed{\leftarrow}\boxed{P}\boxed{+}\boxed{2}\boxed{0}$
$\boxed{\times}\boxed{\alpha}\boxed{\leftarrow}\boxed{S}\boxed{\alpha}\boxed{\rightarrow}\boxed{3}\boxed{4}\boxed{0}\boxed{0}\boxed{\blacktriangleright}\boxed{\text{'}}\boxed{6}\boxed{0}\boxed{\times}\boxed{\alpha}\boxed{\leftarrow}\boxed{H}\boxed{+}\boxed{2}\boxed{5}\boxed{\times}\boxed{\alpha}\boxed{\leftarrow}$
$\boxed{O}\boxed{+}\boxed{5}\boxed{5}\boxed{\times}\boxed{\alpha}\boxed{\leftarrow}\boxed{P}\boxed{+}\boxed{3}\boxed{5}\boxed{\times}\boxed{\alpha}\boxed{\leftarrow}\boxed{S}\boxed{\alpha}\boxed{\rightarrow}\boxed{3}\boxed{1}\boxed{0}\boxed{5}\boxed{0}\boxed{\text{ENTER}}$.

7. Enter the list of variables, in the same order as they are in the constraints and in the objective function: ⬅{ } α⬅H SPC α⬅O SPC α⬅P SPC α⬅S ENTER.

8. Since you wish to minimize the objective function, type **"MIN"** into the final field: →"" α α M I N ENTER.

9. Press ▮OK▮ to compute the solution. After a bit, you will see a message box: Solution found. Press ▮OK▮ again.

Result (to one decimal place):
```
3: { 1.0 4.0 7.0 }
2: [[ -1.0 -1.5 -0.5 0.4 20.0 ]
   [ 1.0 1.0 0.2 -0.2 10.0 ]
   [ -50.0 -110.0 -23.0 17.0 500.0 ]
   [ -0.8 -1.2 -0.3 -0.1 156.0 ]
   [ 0.0 0.0 -1.0 -1.0 0.0 ]
1: { :h: 20.0 :o: 0.0 :p: 0.0 :s: 10.0 }
```

Thus, the minimum cost solution is to buy 20 bales of hay and 10 sacks of sweet feed weekly. (Note, however, that the herd will be eating 500 surplus pounds of roughage a week with this solution— so be prepared for the consequences!)

Try one more example....

Example: Find positive values for the variables x, y, and z that satisfy

$$x + 2y + z \leq 16$$
$$4x + y + 3z \leq 30$$
$$x + 4y + 5z \leq 40$$

and for which the *least* of the three values, x, y and z, is as large as possible.

This kind of problem, called a *MaxMin* problem, requires a small trick—adding a variable some inequalities to the problem. Note that the minimum of the variables x, y, and z is the largest value of the objective value, f, for which $f \leq x$, $f \leq y$, and $f \leq z$. Thus, you must rewrite the set of constraints to include f as a variable and the three new constraints involving f (see step 3 below).

1. *Define the variables.* The variables here are x, y, z, and f.

2. *Find the objective function.* The objective function is simply f.

3. *Find the set of constraints.* The set of constraints, after making the MaxMin additions is:

$$
\begin{aligned}
x + 2y + z \qquad &\leq 16 \\
4x + y + 3z \qquad &\leq 30 \\
x + 4y + 5z \qquad &\leq 40 \\
-x \qquad\quad + f &\leq 0 \\
-y \quad + f &\leq 0 \\
-z + f &\leq 0
\end{aligned}
$$

4. *Enter the LP problem into the* LINPRG *and compute the solution, if one exists.* Begin LINPRG *by typing* $\boxed{\alpha}\,\boxed{\alpha}\,\boxed{\text{L}}\,\boxed{\text{I}}\,\boxed{\text{N}}\,\boxed{\text{P}}\,\boxed{\text{R}}\,\boxed{\text{G}}\,\boxed{\text{ENTER}}$ *or pressing* $\boxed{\text{VAR}}$ *(then* $\boxed{\text{NXT}}$ *or* $\boxed{\leftarrow}\boxed{\text{PREV}}$ *as needed)* **LINPR**.

5. Enter the objective function in the **OBJECTIVE** field, including all variables and the right-hand constant (0 here): $0x + 0y + 0z + f = 0$.
   Press $\boxed{\text{'}}\,\boxed{0}\,\boxed{\times}\,\boxed{\alpha}\,\boxed{\leftarrow}\,\boxed{\text{X}}\,\boxed{+}\,\boxed{0}\,\boxed{\times}\,\boxed{\alpha}\,\boxed{\leftarrow}\,\boxed{\text{Y}}\,\boxed{+}\,\boxed{0}\,\boxed{\times}\,\boxed{\alpha}\,\boxed{\leftarrow}\,\boxed{\text{Z}}\,\boxed{+}\,\boxed{\alpha}\,\boxed{\leftarrow}\,\boxed{\text{F}}$
   $\boxed{\leftarrow}\,\boxed{=}\,\boxed{0}\,\boxed{\text{ENTER}}$.

6. Enter the set of constraints as a list: ⬅{} ' α⬅X + 2 × α⬅
Y + α⬅Z α⬅ 3 1 6 ▶ ' 4 × α⬅X + α⬅Y + 3 ×
α⬅Z α⬅ 3 3 0 ▶ ' α⬅X + 4 × α⬅Y + 5 × α
⬅Z α⬅ 3 4 0 ▶ ' +/− α⬅X + α⬅F α⬅ 3 0 ▶ '
+/− α⬅Y + α⬅F α⬅ 3 0 ▶ ' +/− α⬅Z + α⬅F
α⬅ 3 0 ENTER.

7. Enter the list of variables: ⬅{} α⬅X SPC α⬅Y SPC α⬅Z
SPC α⬅F ENTER.

8. Since you wish to maximize the objective function, leave **"MAX"**
in the final field and press █ OK █ to compute the solution. After a
bit, you will see a message box: Solution found. Now
press █ OK █ once again.

   Result (to three decimal places):
```
3: { 1.000 3.000 7.000 4.000 5.000 2.000 }
2: [[ 0.125  0.375 -0.500  0.125  3.750 ]
   [ 0.125 -0.625  0.500  0.125  3.750 ]
   [ 2.750  1.250 -4.000 -1.250  2.500 ]
   [ 0.125  0.375  0.500  0.125  3.750 ]
   [ 1.500 -0.500 -1.000 -0.500  1.000 ]
   [-0.875  0.375  0.500  0.125  3.750 ]
   [-0.125 -0.375 -0.500 -0.125 -3.750 ]
   [-1.000 -1.000  0.000  0.000  0.000 ]]
1: { :x: 3.750 :y: 3.750 :z: 3.750
     :f: 3.750 }
```

It isn't unusual for a solution to a *MaxMin* problem to yield results
in which the solution variable values are identical to one another.
The value 3.75 represents the largest possible minimum value for x,
y, and z such that the inequalities are satisfied.

Generally, *MaxMin* formulations are useful when you wish to be sure that all
decision variables have as large a value as possible, and that the smallest of the
values is as large as possible. Their counterpart, *MinMax* formulations, are useful
when you wish to be sure that all decision variables have as small a value as
possible, and that the largest of the values is as small as possible.

6. ANALYTIC GEOMETRY

Introduction to Vectors

Analytic geometry is a marriage of algebra and geometry. Geometric concepts such as points, lines, planes, circles and angles are given algebraic descriptions and can thus be analyzed without necessarily portraying them graphically.

Central to this description method is the *vector*. In geometric terms, a vector is a directed line segment. Since it is a segment, it has a finite *length* or *magnitude*— also called its *absolute value*. And the *direction* of a vector is denoted by its two endpoints, the initial endpoint first and the terminal endpoint second. For example, the vector from point A to point B might be referred to as $\vec{AB}$ (whereas the vector from point B to point A is denoted as $\vec{BA}$). Or, if you assume implicitly that the initial point is always the origin (0,0,0), then vectors are especially useful to describe *points*: Every point can be described as a vector of its coordinates.

To illustrate this, use point (-3,7,2). There is a directed line segment—a vector— connecting it to the origin (0,0,0). Its coordinates form a set of instructions about how to get to it starting from the origin: "Move three units in the negative direction along the *x*-axis, then seven units in the positive direction of the *y*-axis and parallel to it, and finally two units in the positive direction of the *z*-axis and parallel to it." The coordinates provide a more algebraic (and analytic) description of the direction of the vector than does the geometric description, $\vec{AB}$. Thus, you can describe the point as a vector: [-3 7 2].

The notation used for vectors is not accidental. They behave algebraically like matrices with one of its dimensions equal to one, so square brackets are used— just as with matrices. Indeed, it is the fact that vectors can make use of the powerful analytic capabilities of matrices that renders them so central to analytic geometry. Algebraically, a matrix is nothing more than a vector of vectors, and each row or each column of a matrix is itself a vector. Accordingly, the HP 48 uses the same delimiters for both matrices and vectors (together called *arrays).

Also, note that any vector in three-dimensional space can be treated as the sum of three *basis* vectors, each running from the origin along one of the coordinate axes. The length of each basis vector is a *component* of the vector, shown in the vector notation explicitly: The vector [4 9 -1], for example, has an *x*-component of 4, a *y*-component of 9 and a *z*-component of −1.

Vector Operations

The basic vector operations—addition, subtraction, and scalar multiplication—work just like their equivalent matrix operations....

Example: Add the two vectors [4 9 -1] and [3 -1 2].

1. Enter the two vectors onto the stack: ⟵[ ] 4 SPC 9 SPC 1 +/− ENTER ⟵[ ] 3 SPC 1 +/− SPC 2 ENTER.
2. Add: +. Result: **[7 8 1]**

Example: Subtract the vector [3 -1 2] from the vector [4 9 -1].

1. Enter the vector [4 9 -1]: ⟵[ ] 4 SPC 9 SPC 1 +/− ENTER.
2. Enter the vector [3 -1 2]: ⟵[ ] 3 SPC 1 +/− SPC 2 ENTER.
3. Subtract: −. Result: **[1 10 -3]**

Example: Multiply the vector [4 9 -1] by the scalar 5.3.

1. Enter the vector [4 9 -1]: ⟵[ ] 4 SPC 9 SPC 1 +/− ENTER.
2. Type in 5.3 and multiply: 5 . 3 ×.
   Result: **[21.2 47.7 -5.3]**

"Multiplying" two vectors is <u>not</u> analogous to arithmetic. There are two kinds of vector products: the *dot product* (or *inner product*) and the *cross product*.

The dot product of two vectors is defined when the two vectors have the same number of elements, n. Thus, given two vectors $r=[r_x \ r_y \ r_z]$ and $s=[s_x \ s_y \ s_z]$, the dot product, $r \cdot s$, is $r_x s_x + r_y s_y + r_z s_z$. As you may know, the HP 48 has a built-in command to compute the dot product.

Example: Find the dot product of [4 9 -1] and [5 -3 2].

1. Enter the first vector: ⟵[ ] 4 SPC 9 SPC 1 +/− ENTER.
2. Enter the second vector: ⟵[ ] 5 SPC 3 +/− SPC 2 ENTER.
3. Compute the dot product: MTH VECTR DOT. Result: **-9**

Note how similar the dot product is to what you do when computing a single element in matrix multiplication: the first vector is treated as a "row," the second as a "column"—and the result is a single number.

By contrast, the *cross product* of two vectors is a third vector—one perpendicular to both of the other vectors (assuming all three vectors originate at the same point: Given two vectors, $\mathbf{r}=[r_x\ r_y\ r_z]$ and $\mathbf{s}=[s_x\ s_y\ s_z]$, their cross product, $\mathbf{r \times s}$, is the vector $[\ r_y s_z - r_z s_y \quad r_z s_x - r_x s_z \quad r_x s_y - r_y s_x\ ]$.

Example: Find the cross product of [4 9 -1] and [5 -3 2].

    1. Enter the first vector: ⟵ [ ] 4 SPC 9 SPC 1 +/− ENTER.

    2. Enter the second vector: ⟵ [ ] 5 SPC 3 +/− SPC 2 ENTER.

    3. Compute the cross product: CROSS.

            Result: [15 -13 -57]

Like matrix multiplication, the order in which you perform the cross product is important. Look at this diagram:

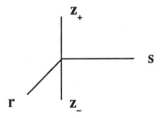

The point is this: When taking the cross product $\mathbf{r \times s}$, you will get the $\mathbf{z}_+$ vector; when taking the other cross product, $\mathbf{s \times r}$, you will get the $\mathbf{z}_-$ vector.

Example: Find the cross product of [5 -3 2] with [4 9 -1].

    1. Enter the first vector: ⟵ [ ] 5 SPC 3 +/− SPC 2 ENTER.

    2. Enter the second vector: ⟵ [ ] 4 SPC 9 SPC 1 +/− ENTER.

    3. Compute the cross product: CROSS.

            Result: [-15 13 57]

    Note that the result is the negative of the previous result.

Finding Angle and Magnitude of Vectors

A vector has both magnitude (length) and direction. It should therefore be possible to find these parameters easily for a vector entered in standard form.

Example: Find the length of the vector [4 9 -1].

    1. Enter the vector onto the stack: ⇦[ ] 4 SPC 9 SPC 1 +/– ENTER.

    2. Find its length: MTH ▉VECTR▉ ▉ABS▉. Result: 9.89949493661

Finding the "direction" of a vector is more complicated. To determine an angle or direction, you must first decide the reference direction against which you are measuring the angle. For a vector in three dimensions, you use the three coordinate axes as your three reference directions; the vector forms a different angle with respect to each axis. The three direction angles for a vector can be computed from the vectors com-ponents and its length:

$$\theta_x = \cos^{-1}\frac{v_x}{|V|} \qquad \theta_y = \cos^{-1}\frac{v_y}{|V|} \qquad \theta_z = \cos^{-1}\frac{v_z}{|V|}$$

where v_x, v_y, and v_z are the vector components of the vector V.

Example: Find the direction angles of the vector [4 9 -1].

    1. Assuming that you're in Degree mode (press ⇦RAD, if necessary) and that the result of the previous example is still sitting on level 1, make two copies of the vector's magnitude: ENTER ENTER.

    2. Compute the x-direction angle: 4 ENTER SWAP / ⇦ACOS.
        Result: 66.1677009381

    3. Rotate another copy of the magnitude into level 1 and compute the y-direction angle: ⇦STACK ▉ROT▉ 9 ENTER SWAP / ⇦ACOS.
        Result: $24.613597653.$

    4. Repeat step 3 using the z-component: ▉ROT▉ 1 +/– ENTER SWAP / ⇦ACOS. Result: 95.7976363295

The computation of the direction angles for a vector can be easily automated with a short program, which is what VDIR does (see page 317 for listing).

Example: Use VDIR to find the direction angles for the vector [-5 3 2].

1. Put the vector onto the stack: ⬅[｜⟨⟩｜] 5 +/- SPC 3 SPC 2 ENTER.
2. Execute VDIR: α α V D I R ENTER or VAR (then NXT or ⬅ PREV as needed) VDIR.

   Result: { 144.204240085 60.8784319297
   71.0681768192 } (in Deg mode).

Overview of Analytical Geometry Examples

This part of this chapter organizes topics and examples in analytical geometry according to the information you are given. For instance, the section titled "Given: Two Points" shows examples of computations you (and your HP 48) can perform if you already know the coordinates of two points. And so on. The examples in this part are organized as follows:

Given: Two Points

- Find the distance between them.
- Find the equation of the line they determine.
- Find the midpoint of the line segment they determine.
- Find the coordinate of the point on that line segment that divides it into a given ratio.
- Find the equation of the perpendicular bisector of that line segment.

Given: Three Points

- Determine if they are collinear.
- Find the equations of the lines they determine.
- Find the equation of the plane they determine.
- Find the equation of the perpendicular containing one point to the line containing the other two points.
- Find the distance of one point from the line containing the other two points.
- Find the area of the triangle they determine.
- Find the coordinates of the centroid of the triangle they determine.

Given: A Line or Point-and-Slope

- Find the desired alternative equation (general, intercept, or parametric) for a line.

Given: A Point and a Line

- Determine if the point is collinear with the line.
- Find the equation of the line perpendicular through a point on the line.
- Find the equation of the line perpendicular through a point not on the line.
- Find equation of the line parallel through a point not on the line.
- Find the distance from the point to the line.
- Find the equation of the plane they determine.
- Find the equation of the plane from a normal and a point in the plane.

Given: Two Lines

- Determine if they are parallel, skew, concurrent, collinear, perpendicular.
- Determine the point of intersection.
- Find the distance between two parallel lines.
- Find the angle formed by their intersection.
- Find the plane they determine.
- Find the plane from its traces (three lines).
- Find the line perpendicular to plane they determine (cross product).

Given: A Point and a Plane

- Find the distance of the point to the plane.
- Find the equation of the plane from a parallel plane and a point in the plane.
- Find the equation of the plane from a perpendicular plane and a point in the plane.

Given: One or Two Planes

- Find the equation of the line of intersection.
- Find the angle between the planes.
- Find the traces of a plane.

Given: Two Points

Distance Between Points

The distance between two points, $[x_1 \, y_1 \, z_1]$ and $[x_2 \, y_2 \, z_2]$, is given by the following formula $d = \sqrt{(x_2 - x_1)^2 + (y_2 - y_1)^2 + (z_2 - z_1)^2}$.

While you can compute the distance by using this formula, note that simply subtracting the vector form of one point from another gives the vector connecting them. Then you only need to find the length (absolute value) of this difference vector to compute the distance between points.

Example: Find the distance between the two points, [2 4 -6] and [-1 -2 3].

    1. Enter the second point (as a vector): ⇦ [ ] 1 +/− SPC 2 +/− SPC 3 ENTER.

    2. Enter the first point likewise: ⇦ [ ] 2 SPC 4 SPC 6 +/− ENTER.

    3. Compute the distance between the points: − MTH VECTR ABS.

<u>Result</u>: 11.2249721603

Midpoints

The coordinates of the midpoint are simply the "average" of the two points. In vector terms, this means you must add the points together and divide by two.

Example: Find the midpoint of the line segment between the points [2 4 -6] and [-1 -2 3].

1. Enter the second point (as a vector): ⏴ [ ] 1 +/− SPC 2 +/− SPC 3 ENTER.
2. Enter the first point likewise: ⏴ [ ] 2 SPC 4 SPC 6 +/− ENTER.
3. Compute the midpoint: + 2 ÷. <u>Result</u>: **[.5 1 -1.5]**

The coordinates of a point P_3 which divides a segment into a given ratio $m{:}n$ can be individually computed as follows:

$$x_3 = \frac{nx_1 + mx_2}{m+n} \qquad y_3 = \frac{ny_1 + my_2}{m+n} \qquad z_3 = \frac{nz_1 + mz_2}{m+n}$$

This is a kind of weighted average of the coordinates of the endpoints. Of course, using vectors, you can compute all of the coordinates simultaneously, as the next example demonstrates.

Example: Find the coordinates of the point on the line segment between the points, [2 4 -6] and [-1 -2 3], that divides the segment into a 3:2 ratio.

1. Enter the first point: ⏴ [ ] 2 SPC 4 SPC 6 +/− ENTER.
2. Multiply it by the fractional weighting for the first point ($n/(m+n)$): 2 ENTER 5 ÷ ×.
3. Enter the second point: ⏴ [ ] 1 +/− SPC 2 +/− SPC 3 ENTER.
4. Multiply it by the fractional weighting for the second point ($m/(m+n)$): 3 ENTER 5 ÷ ×.
5. Add the two weighted vectors to find the coordinates of the desired point: +. <u>Result</u>: **[.2 .4 -.6]**

Lines

You may also find the equations of particular lines associated with the two points:

- The line containing both points. (Note that there are several different forms of equations for a line. See pages 193-195 for examples of converting between different forms.)

- The perpendicular bisector of the line segment linking the two points.

The next few examples illustrate each of these computations.

Example: Find the equation of the line in the xy-plane containing the points [2 4] and [-1 3].

1. Enter the first point: ⟨←⟩⟨[ ]⟩⟨2⟩⟨SPC⟩⟨4⟩⟨ENTER⟩.

2. Enter the second point: ⟨←⟩⟨[ ]⟩⟨1⟩⟨+/−⟩⟨SPC⟩⟨3⟩⟨ENTER⟩.

3. Compute the slope of the line: ⟨←⟩⟨STACK⟩ **OVER** ⟨−⟩⟨PRG⟩ **LIST** **OBJ→** ⟨◀⟩⟨SWAP⟩⟨÷⟩. <u>Result</u>: .333333333333 (or 1/3).

4. Compute the y-intercept of the line: ⟨SWAP⟩ **OBJ→** ⟨◀⟩⟨SWAP⟩⟨←⟩ ⟨STACK⟩ **ROT** ⟨×⟩⟨+/−⟩⟨+⟩. <u>Result</u>: 3.33333333333 (or 10/3).

Thus the equation of the line is $y = \dfrac{1}{3}x + \dfrac{10}{3}$.

Of course, the procedure in the previous example can be easily automated. The program, P2→L, listed on page 297, takes the two points in vector form from the stack and returns the slope-intercept form of the equation for the line.

Example: Repeat the previous example using the P2→L program.

1. Enter the first point: ⟨←⟩⟨[ ]⟩⟨2⟩⟨SPC⟩⟨4⟩⟨ENTER⟩.

2. Enter the second point: ⟨←⟩⟨[ ]⟩⟨1⟩⟨+/−⟩⟨SPC⟩⟨3⟩⟨ENTER⟩.

3. Execute P2→L: ⟨α⟩⟨α⟩⟨P⟩⟨2⟩⟨→⟩⟨→⟩⟨L⟩⟨ENTER⟩ or ⟨VAR⟩ (then ⟨NXT⟩ or ⟨←⟩⟨PREV⟩ as needed) **P2→L**. <u>Result</u>: 'y=1/3*x+10/3'

So far, all the lines you have seen have been limited to the xy-plane; the points used to determine them have had just two coordinates (i.e. the z-coordinate was zero). But lines exist in three-dimensional space. How do you express their equations?

The best method is parametric description—describing in three short equations how each of the three components change with an independent parameter. For example, the following set of equations describes a line in space, parametrically:

$$x = t + 2, \quad y = 2t - 3, \quad z = 2t + 4$$

You can tell that this set of equations represents a line because all three are *linear* in the parameter, t—i.e. each coordinate changes in a straight line as t changes.

You can find the *parametric* form for the equation of a line in space from two points, $[\, x_1\, y_1\, z_1\,]$ and $[\, x_2\, y_2\, z_2\,]$, as follows:

1. Find the vector $[\, A\ B\ C]$ of the line segment connecting the two points.

2. Create the parametric equations $x = At + x_1, \quad y = Bt + y_1, \quad z = Ct + z_1$

Example: Find the set of parametric equations for the line determined by the points $[\, 2\ 4\ \text{-}6\,]$ and $[\, \text{-}1\ \text{-}2\ 3\,]$.

1. Enter the first point onto the stack and make an extra copy: ⬅️ [ ] ② SPC ④ SPC ⑥ +/– ENTER ENTER.

2. Enter the second point: ⬅️ [ ] ① +/– SPC ② +/– SPC ③ ENTER.

3. Find the vector connecting the points: ⊟. <u>Result</u>: **[3 6 -9]**

4. Create the parametric equations. You can either assemble equations manually, using the first point and the computed vector; or you can use the program **PD→P** (see page 291), which takes a point (in vector form) from level 2 and a vector representing the line from level 1 and assembles the proper parametric equations for the line: α α P D ➡️ → P ENTER or VAR (then NXT or ⬅️ PREV as needed) **PD→P**.
   <u>Result</u>: **{ 'x=3*t+2' 'y=6*t+4' 'z=-(9*t)-6' }**

Another line that is determined by two points is the perpendicular bisector of the line segment that joins the two points. Because it is perpendicular, it will have a slope that is the negative reciprocal of the slope of the line segment. Because it is the bisector, it contains the midpoint of the line segment. Thus, if the slope of the line segment is m and the coordinates of the midpoint is [a b], then the equation of the perpendicular bisector is: $y = -\dfrac{x}{m} + \left(\dfrac{a}{m} + b\right)$.

Example: Find the perpendicular bisector for the segment whose endpoints are [2 4] and [-1 3].

1. Enter the first point: ⇐[ ] 2 SPC 4 ENTER.

2. Enter the second point: ⇐[ ] 1 +/− SPC 3 ENTER.

3. Make copies of the two points and compute the slope of the perpendicular bisector: ⇐STACK NXT **DUP2** SWAP −PRG **LIST OBJ→** ◄SWAP ÷ 1/x +/−. Result: ‾3

4. Compute the coordinates of the midpoint: SWAP ⇐STACK **ROT** + 2 ÷. Result: [.5 3.5]

5. Now find the y-intercept of the perpendicular bisector: PRG **LIST OBJ→** ◄SWAP ⇐STACK **ROT** × +/− +.
   Result: 5

Thus the equation of the perpendicular bisector is $y = -3x + 5$.

The short program, P2→PB (see listing on page 297), automates the process of determining the equation of the perpendicular bisector....

Example: Repeat the previous example using the P2→PB program.

1. Enter the first point: ⇐[ ] 2 SPC 4 ENTER.

2. Enter the second point: ⇐[ ] 1 +/− SPC 3 ENTER.

3. Execute P2→PB: α α P 2 ⇒ → P B ENTER or VAR (then NXT or ⇐PREV as needed) **P2→PB**. Result: 'y=-(3*x)+5'

Finally, here's how you compute the perpendicular bisector for a line in space.

Example: Find the perpendicular bisector of the line segment connecting the points [2 4 -6] and [-1 -2 3].

1. Enter the two points: ⟵[[ ]][2][SPC][4][SPC][6][+/−][ENTER]⟵[[ ]][1] [+/−][SPC][2][+/−][SPC][3][ENTER].

2. Make copies of the two points, compute the vector connecting them, and find its negative: ⟵[STACK][NXT] **DUP2** [−][+/−].

   Result: **[−3 −6 9]**

   This is the direction vector for the perpendicular bisector.

3. Find the coordinates of the midpoint: [SWAP][NXT] **ROT** [+][2][÷].

   Result: **[.5 1 −1.5]**

   This is a point on the perpendicular bisector.

4. Assemble the set of parametric equations representing the perpendicular bisector: [SWAP][VAR] **PO→P**.

   Result: **{ 'x=-(3*t)+1/2' 'y=-(6*t)+1'**
   'z=9*t-3/2' }

Remember, too, that you can always use this approach to find the parametric equations for points in the xy-plane, simply by including a zero as the z-component.

Given: Three Points

Whenever you have three distinct points, you need to know whether they are *collinear* (i.e. all contained in a single line). If they are *noncollinear*, then they determine a plane, a triangle, and three distinct intersecting lines. If they are collinear, the three points only determine a single line.

Example: Determine if the points [4 2 -6], [5 -3 2], [-1 2 3] are collinear.

1. Enter the points: ⤆[ ] 4 SPC 2 SPC 6 +/− ENTER ⤆[ ] 5 SPC 3 +/− SPC 2 ENTER ⤆[ ] 1 +/− SPC 2 SPC 3 ENTER.

2. Compute the vector representing the segment between the second and third points: −. <u>Result:</u> **[6 −5 −1]**

3. Divide each component of the result into the corresponding component of the first point: 4 ENTER 6 ÷ 2 ENTER 5 +/− ÷ 6 +/− ENTER 1 +/− ÷. <u>Results:</u> **.6667 −.4000 6.0000**

4. Compare the ratios: Since the ratios are not equal for all components, then the three points are *non*collinear.

The program COLIN? (see page 278) makes it easier to test for collinearity (and is useful in other programs). It tests whether a point (in vector form) on level 1 is collinear with the vector on level 2. If so, it returns a 1; if not, it returns a 0.

Example: Repeat the previous example using COLIN?.

1. Enter the first two points onto the stack: ⤆[ ] 4 SPC 2 SPC 6 +/− ENTER ⤆[ ] 5 SPC 3 +/− SPC 2 ENTER.

2. Compute the vector connecting the two points: −.

3. Enter the third point: ⤆[ ] 1 +/− SPC 2 SPC 3 ENTER.

4. Execute the COLIN? test: α α C O L I N ⤆ ◄ ENTER or VAR (then NXT or ⤆ PREV as needed) **COLIN**.

   <u>Result:</u> **0**   *Noncollinear*

Lines and Planes

If three points are noncollinear, they determine three lines. You may find these three lines easily by using the P2→L program for each combination of points.

Example: Find the equations of the three lines determined by the noncollinear points [4 6], [-2 1] and [5 -2].

1. Enter the first two points and execute P2→L: ⬅[[ ]]④SPC⑥ENTER ⬅[[ ]]②+/- SPC①ENTER VAR `P2+L`.

   Result: $'y=5/6*x+8/3'$

2. Enter the second and third points and execute P2→L: ⬅[[ ]]②+/- SPC①ENTER ⬅[[ ]]⑤SPC②+/-ENTER VAR `P2+L`.

   Result: $'y=-(3/7*x)+1/7'$

3. Enter the first and third points and execute P2→L: ⬅[[ ]]④SPC⑥ ENTER ⬅[[ ]]⑤SPC②+/-ENTER VAR `P2+L`.

   Result: $'y=-(8*x)+38'$

Of course, three noncollinear points also determine a unique plane. The general form of the equation for a plane is $Ax + By + Cz + D = 0$.

It turns out that the coefficients A, B, and C determine the "orientation" of the plane and are equal to components of the vector perpendicular (or *normal*) to the plane. The D coefficient identifies the particular plane (from the infinitely large set of parallel planes with the given orientation) determined by the three points.

Thus, to compute the equation of the plane, you must find a line perpendicular to at least two of the lines determined by the points. This is easily accomplished by finding the cross product of two of the vectors representing the line segments determined by the points. Once you have the orientation of the plane, you need only to substitute the coordinates of any one of the points to determine the D coefficient. This substitution is efficient accomplished using the dot product of the normal vector and the point. The next example illustrates the full procedure....

Example: Find the equation of the plane determined by the three noncollinear points, [4 2 -6], [5 -3 2], [-1 2 3].

1. Enter the first two points and compute the vector of the line segment connecting them: ⬅[] 4 SPC 2 SPC 6 +/− ENTER ⬅[] 5 SPC 3 +/− SPC 2 ENTER −.

2. Enter the first and third points and compute the vector of the line segment connecting them: ⬅[] 4 SPC 2 SPC 6 +/− ENTER ⬅[] 1 +/− SPC 2 SPC 3 ENTER −.

3. Compute the cross product of the two line segment vectors: MTH **VECTR** **CROSS**. Result: [−45 −49 −25]

   The components of this normal vector are the coefficients A, B, and C respectively in the equation for the plane.

4. Enter the first point again and compute the negative of the dot product of the normal vector and the first point (you should get the same result using any of the points): ⬅[] 4 SPC 2 SPC 6 +/− ENTER **DOT** +/−. Result: 128. This is the coefficient D in the equation of the plane, so the equation is $-45x - 49y - 25z + 128 = 0$.

A related task is to find the equation of the line containing one of the points that is perpendicular to the line determined by the remaining two points.

Example: Find the equation of a line containing the point [4 6] perpendicular to the line determined by the points [-1 5] and [3 -5].

1. Enter the two points of the line: ⬅[] 1 +/− SPC 5 ENTER ⬅[] 3 SPC 5 +/− ENTER.

2. Find the slope of the perpendicular: SWAP − PRG **LIST** **OBJ→** ⬅ ÷ +/−. Result: .4   (or 2/5)

3. Enter the point on the perpendicular: ⬅[] 4 SPC 6 ENTER.

4. Find the y-intercept of the perpendicular: **OBJ→** ⬅ SWAP ⬅ STACK **ROT** × +/− +. Result: 4.4. Thus the equation of the perpendicular through [4 6] is $y = \dfrac{2}{5}x + \dfrac{22}{5}$.

The distance between point [r s] and line $Ax + By + C = 0$, is $d = \dfrac{|Ar + Bs + C|}{\sqrt{A^2 + B^2}}$

Example: Find the distance from the point [4 6] to the line determined by the points [-1 5] and [3 -5].

1. Enter the two points of the line: ⦗←⦘⦗[ ]⦘⦗1⦘⦗+/−⦘⦗SPC⦘⦗5⦘⦗ENTER⦘⦗←⦘⦗[ ]⦘ ⦗3⦘⦗SPC⦘⦗5⦘⦗+/−⦘⦗ENTER⦘.

2. Find the general equation of the line. The quickest method uses the programs **P2→L** and **I→GEN** (see page 284): ⦗VAR⦘ **P2→L** **I→GEN**.

    <u>Result:</u> 2: [-2.5 -1 2.5]
    1: '5/2-5/2*x-y=0'

3. Drop the symbolic form of the line, make an extra copy and remove the last element from the vector: ⦗←⦘⦗ENTER⦘⦗MTH⦘ **MATR** **COL** ⦗3⦘ **COL−** ⦗←⦘.

4. Append a 1 as the third element of the vector of the distant point and enter the result ([4 6 1]): ⦗←⦘⦗[ ]⦘⦗4⦘⦗SPC⦘⦗6⦘⦗SPC⦘⦗1⦘⦗ENTER⦘.

5. Compute the distance: ⦗←⦘⦗STACK⦘ **ROT** ⦗MTH⦘ **VECTR** **DOT** **ABS** ⦗SWAP⦘ **ABS** ⦗÷⦘.

    <u>Result:</u> 5.01377413078

The short program **DtoL** (see page 280) takes a point in vector form on level 2 and an array on level 1. The array is a matrix containing the two points that determine a line. The program returns the distance from the point to the line.

Example: Repeat the previous example using the **DtoL** program.

1. Enter the distant point ([4 6]): ⦗←⦘⦗[ ]⦘⦗4⦘⦗SPC⦘⦗6⦘⦗ENTER⦘.

2. Enter a matrix with the points representing a line: ⦗→⦘⦗MATRIX⦘⦗1⦘ ⦗+/−⦘⦗ENTER⦘⦗5⦘⦗ENTER⦘⦗▼⦘⦗3⦘⦗ENTER⦘⦗5⦘⦗+/−⦘⦗ENTER⦘⦗ENTER⦘.

3. Execute **DtoL**: ⦗α⦘⦗α⦘⦗D⦘⦗←⦘⦗T⦘⦗←⦘⦗O⦘⦗L⦘⦗ENTER⦘ or ⦗VAR⦘ (then ⦗NXT⦘ or ⦗←⦘⦗PREV⦘ as needed) **DTOL**. <u>Result:</u> 5.01377413078

Triangles

The other geometric form defined by three points is the triangle. These examples show how to find the perimeter, area, median length and centroid of the triangle.

Example: Find the perimeter of the triangle formed by the three points [4 6], [-1 5], and [3 -5].

1. Enter the first two points and compute the length of the segment connecting them: ⟵ [ ] 4 SPC 6 ENTER ⟵ [ ] 1 +/− SPC 5 ENTER − MTH **VECTR** **ABS**.

2. Enter the first and third points and compute the length of the segment connecting them: ⟵ [ ] 4 SPC 6 ENTER ⟵ [ ] 3 SPC 5 +/− ENTER − **ABS**.

3. Enter the second and third points and compute the length of the segment connecting them: ⟵ [ ] 1 +/− SPC 5 ENTER ⟵ [ ] 3 SPC 5 +/− − **ABS**.

4. Compute the perimeter: + +. <u>Result:</u> 26.9147101451

The area of a triangle determined by three points can be found by using the *length* of cross product of two of the vectors determined by the three points. If **r**, **s**, and **t** are the three vectors, then the area of the triangle is given by

$$Area = \frac{1}{2}|r \times s| = \frac{1}{2}|r \times t| = \frac{1}{2}|s \times t|$$

Note that the direction (or sign) of the cross product doesn't matter for computing the area of the determined triangle because you are only interested its length.

Example: Find the area of the triangle formed by [4 6 -2], [-1 5 3], [3 -5 1].

1. Compute a vector formed by the first and second points: ⟵ [ ] 4 SPC 6 SPC 2 +/− ENTER ⟵ [ ] 1 +/− SPC 5 SPC 3 ENTER −.

2. Compute a vector formed by the first and third points: ⟵ [ ] 4 SPC 6 SPC 2 +/− ENTER ⟵ [ ] 3 SPC 5 +/− SPC 1 ENTER −.

3. Compute the area: **CROSS** **ABS** 2 ÷. <u>Result:</u> 37.8153408024

A *median* of a triangle is a line segment connecting a vertex with the midpoint of the opposite side.

Example: Find the length of the median of a triangle from the vertex [4 6] to the midpoint of the segment with endpoints [-1 5] and [3 -5].

1. Find the midpoint of the side opposite the vertex [4 6]: ⬅[]1
   +/− SPC 5 ENTER ⬅[]3 SPC 5 +/− ENTER + 2 ÷.

2. Enter the vertex endpoint of the median: ⬅[]4 SPC 6 ENTER.

3. Compute the length of the median: − **ABS** .

   Result: **6.7082039325**

The *centroid* of a triangle is the point where the three medians intersect. It divides each median so that the distance from the centroid to the vertex is twice the distance from the centroid to the midpoint of the opposite side. The coordinates of the centroid are the average of the coordinates of the three sides.

Example: Find the coordinates of the centroid of the triangle determined by the points [4 6], [-1 5], and [3 -5].

1. Enter the points on the stack: ⬅[]4 SPC 6 ENTER ⬅[]1 +/−
   SPC 5 ENTER ⬅[]3 SPC 5 +/− ENTER.

2. Compute the coordinates of the centroid: + + 3 ÷.

   Result: **[2 2]**

Given: A Line or Point-and-Slope

As you know, with two points, you may determine the equation of the line that contains them. But you can also find the equation of the line if you know just a single point on the line and line's slope: Given the point [r s] in the xy-plane and a slope m, the equation of the line (in slope-intercept form) is $y = mx + (s - mr)$.

Example: Find the equation of a line with slope 4, containing the point [3 -8].

1. Compute the y-intercept, $s - mr$: ⑧[+/−][ENTER]④[ENTER]③[×][−].
   Result: ‾20. So the equation in slope-intercept form is $y = 4x - 20$.

For lines in space, the concept of "slope" becomes "direction vector." Thus, to find the equation of a line in space, you need to know only a point on the line and its direction vector. The program PD→P (see page 291) does exactly this—take a point (in vector form) from level 2 and a direction vector from level 1 and computes the set of parametric equations describing the line determined.

Example: Find the equation of a line in space containing the point [3 -8 2] whose direction vector is [2 -1 -2].

1. Enter the point: [←][[ ]]③[SPC]⑧[+/−][SPC]②[ENTER].
2. Then the direction vector: [←][[ ]]②[SPC]①[+/−][SPC]②[+/−][ENTER].
3. Execute PD→P: [VAR] **PD→P** .
   Result: { 'x=2*t+3' 'y=-t-8' 'z=-(2*t)+2' }

There are five important forms of linear equations. Two of them apply only to lines in the xy-plane; three apply to any line, including lines in space:

- *Slope-Intercept form* (also known as *direction* form): $y = mx + b$, where m is the slope and b is the y-intercept. This form only applies to lines in the xy-plane (there is no z-component).

- *General form*: $Ax + By + C = 0$ where, if $C = -AB$, then A is the y-intercept and B is the x-intercept. This form only applies to lines in the xy-plane (there is no z-component).

- *Position-Direction form*: $[x_0 \, y_0 \, z_0]$, $[a \, b \, c]$ where $[x_0 \, y_0 \, z_0]$ is the position vector representing a point on the line and $[a \, b \, c]$ is a direction vector for the line. This form is a vector version of the slope-intercept form and can apply to any line.

- *Parametric form*: $x = pt + x_0$, $y = qt + y_0$, $z = rt + z_0$, where $[x_0 \, y_0 \, z_0]$ is a point on the line and $[p \, q \, r]$ is a direction vector for the line. This form applies to all lines. For lines in the xy-plane, the third equation reduces to $z = 0$. Note that $[x_0{+}p \, \ y_0{+}q \, \ z_0{+}r]$ is a second point on the line.

- *Array form*: $\begin{bmatrix} x_1 & y_1 & z_1 \\ x_2 & y_2 & z_2 \end{bmatrix}$ where $[x_1 \, y_1 \, z_1]$ and $[x_2 \, y_2 \, z_2]$ are points on the line. This form applies to all lines. For lines in the xy-plane, the array only has two columns (there is no z-column). The array form can be generalized to represent any arbitrary collection of points—very useful for performing transformations (as you will see later in this chapter).

Depending on circumstances, one of these forms will be more convenient to find than the others. However, you may find instances when you need to convert from one form to another. Some small programs make these conversions easier:

- PAR→I (see page 290) converts a set of parametric equations to an equation in slope-intercept form (but only possible for lines in the xy-plane).

- P→PD (see page 295) converts a set of parametric equations to a point (in vector form) and a direction vector (position-direction form).

- G→A (see page 282) converts an equation in general form to an array.

- I→GEN (see page 284) converts an equation in slope-intercept form to an equation in general form.

Example: Convert these to slope-intercept form: $x = 3t - 4$, $y = -2t + 5$

1. Enter the set of parametric equations as a list: ⇦{}' α⇦X⇦ =3X α⇦T−4▶' α⇦Y⇦=2+/−X α⇦T+5 ENTER. Note that the parameter must always be t (lower-case), and the other variables x and y.

2. Execute **PAR→I** : α α P A R→I ENTER or VAR (then NXT or ⇦ PREV as needed) **PAR→I**. Result: `'y=7/3-2/3*x'`

Example: Convert the line, $3x - 2y + 5 = 0$, to a set of parametric equations.

1. Enter the line (which is in general form): '3X α⇦X−2X α⇦Y+5⇦=0 ENTER.

2. Convert it to an array: VAR **G→A** . Result:
   ```
   [[ 1  4 ]
    [ 2 5.5 ]]
   ```

3. Now disassemble the array into its two point vectors: MTH **MATR** **ROW** **→ROW** ⇦.

4. Make a copy of one of the points and compute the vector for the line: ⇦STACK **OVER** −.

5. Convert the point and vector on the stack to a set of parametric equations: VAR **PD→P** . Result: { `'x=t+1'` `'y=3/2*t+4'` }

Example: Convert the following set of parametric equations to a line in array form: $x = 3t - 4$, $y = -2t + 5$, $z = t + 3$.

1. Enter the equations: ⇦{}' α⇦X⇦=3X α⇦T−4▶ ' α⇦Y⇦=2+/−X α⇦T+5▶' α⇦Z⇦=α⇦ T+3 ENTER.

2. Convert it to a point and a direction vector: VAR **P→PD** .

3. Make a copy of the point and add it to the vector to form a second point: ⇦STACK **OVER** +.

4. Now assemble the two points into a matrix: 2 MTH **MATR** **ROW** **ROW→** . Result:
   ```
   [[ -4  5  3 ]
    [ -1  3  4 ]]
   ```

Example: Convert the following set of parametric equations to the general form: $x = -5t + 2$, $y = 3t + 1$.

1. Enter the parametric equations: ⏴{} ' α⏴X ⏴= 5 +/- × α⏴T + 2 ▶ ' α⏴Y ⏴= 3 × α⏴T + 1 ENTER.

2. Convert to the slope-intercept form: VAR PAR⊹I.

   Result: `'y=11/5-3/5*x'`

3. Convert to the general form: I⊹GE.

   Result: 2: `[ -.6 -1 2.2 ]`
            1: `'11/5-3/5*x-y=0'`

Example: Convert the array $\begin{bmatrix} -4 & 3 \\ 1 & -2 \end{bmatrix}$ to a line in general form.

1. Enter the array: ➡MATRIX 4 +/- ENTER 3 ENTER ▼ 1 ENTER 2 +/- ENTER ENTER.

2. Now disassemble the array into its component points: MTH MATR ROW ⊹ROW ⏴.

3. Compute the slope-intercept form: VAR P2⊹L.

4. Convert the slope-intercept form to general form: I⊹GE.

   Result: 2: `[ -1 -1 -1 ]`
            1:     `'1-x-y=0'`

Example: Convert the line, $y = 4x - 6$, to array form.

1. Enter the line (in slope-intercept form): ' α⏴Y ⏴= 4 × α ⏴X ⏴- 6 ENTER.

2. Convert the line to general form: VAR I⊹GE SWAP ⏴.

3. Convert to an array: G⊹A. Result: `[[ 1 -2 ]`
                                      `[ 2  2 ]]`

Given: A Point and a Line

The procedures for using a point and a line are very similar to those for using three points. After all, once you find the line determined by two of the three points, the conditions for the two situations are identical. However, there are some practical differences that arise, depending on the form which the linear equation takes. The examples below illustrate this.

Example: Is the point [4 -1] collinear with the line $-2x + y - 4 = 0$?

1. Enter the line: [']2[+/-][×][α][←][X][+][α][←][Y][-][4][←][=][0] [ENTER].

2. Next, convert the line to a direction vector: [VAR] `G→A` [MTH] `MATR` `ROW` `→ROW` [←][-]. Result: `[ -1 -2 ]`

3. Enter the point: [←][[ ]][4][SPC][1][+/-][ENTER].

4. Execute the test for collinearity, `COLIN?`: [VAR] `COLIN`.

   Result: `0`. The point and line are noncollinear.

Example: Find the equation of the plane (in general form) determined by the point [4 -1 2] and the line $x = 3t - 2$, $y = -2t + 5$, $z = t - 3$.

1. Enter the list of parametric equations: [←][{ }]['][α][←][X] [←][=] [3] [×][α][←][T][-][2][▶]['][α][←][Y][←][=][2][+/-][×][α][←][T][+][5][▶] ['][α][←][Z][←][=][α][←][T][-][3] [ENTER].

2. Find the direction vector for the line: [VAR] `P→PD` .

3. Compute the set of coefficients $[A\,B\,C]$ for the equation of the plane: [←][STACK] `OVER` [+][MTH] `VECTR` `CROSS` .

   Result: `[ -1 -7 -11 ]`

4. Enter the point and compute the D coefficient: [←][[ ]][4][SPC][1][+/-] [SPC][2][ENTER] `DOT` [+/-]. Result: `19`

   Thus the equation for the plane is $-x - 7y - 11z + 19 = 0$.

Example: Find the set of parametric equations of the perpendicular to the line $x = 3t - 2$, $y = -2t + 5$, $z = t - 3$, that contains the point [4 -1 2].

1. Enter the list of parametric equations: ⟵[{][}]['][α][⟵][X][⟵][=][3][X] [α][⟵][T][−][2][▶]['][α][⟵][Y][⟵][=][2][+/−][X][α][⟵][T][+][5][▶] ['][α][⟵][Z][⟵][=][α][⟵][T][−][3][ENTER].

2. Convert to a point and direction vector for the line: [VAR] **P→PD** .

3. Compute the direction vector for the perpendicular: [⟵][STACK] **OVER** [+][MTH] **VECTR** **CROSS**. Result: [−1 −7 −11]

4. Enter the point and find the set of parametric equations for the perpendicular: [⟵][[][]][4][SPC][1][+/−][SPC][2][ENTER][SWAP][VAR] **PD→P** .

   Result: { 'x=-t+4' 'y=-(7*t)-1' 'z=-(11*t)+2' }

Two parallel lines, given in point-direction vector form, will have equal direction vectors but different sets of points; the points of one of the lines are noncollinear with the points of the other. If they were collinear, the two lines would then be *concurrent*—essentially the same line. It is relatively easy to find the equation of a line parallel to a given line, through a given point not on that line.

Example: Find the equation of the line parallel to the line $x = 3t - 2$, $y = -2t + 5$, $z = t - 3$, that contains the point [4 -1 2].

1. Enter the list of parametric equations: ⟵[{][}]['][α][⟵][X][⟵][=][3][X] [α][⟵][T][−][2][▶]['][α][⟵][Y][⟵][=][2][+/−][X][α][⟵][T][+][5][▶] ['][α][⟵][Z][⟵][=][α][⟵][T][−][3][ENTER].

2. Convert to point-direction vector form: [VAR] **P→PD** .

3. Replace the point vector for the given line with the given point: [SWAP][◀][⟵][[][]][4][SPC][1][+/−][SPC][2][ENTER][SWAP] .

4. Compute the set of parametric equations for the parallel line: [VAR] **PD→P** . Result: { 'x=3*t+4' 'y=-(2*t)-1' 'z=t+2' }

Finding the distance between a point in space and a line is a bit tricky, because it would seem that you must compute the point of intersection between the given line and the perpendicular containing the given point. However, recall that the given point and any two points on the given line form a triangle.

The area of the triangle is $A = \dfrac{1}{2}bh$ where b is the length of the base and h is the length of the height. If you choose the base of the triangle to run along the given line, then the height is the distance between the given point and the given line. Thus, since you can use the cross product to find the area of a triangle in space, and you can choose any base points and compute the distance between them, you can find the height without computing any coordinates of intersection points.

Example: Find the distance from the point [-5 2 1] to the line defined by $x = 3t - 2$, $y = -2t + 5$, $z = t - 3$.

1. Enter the list of parametric equations: ⬅{}{} ' α ⬅X ⬅= 3 X α ⬅T − 2 ▶ ' α ⬅Y ⬅= 2 +/− X α ⬅T + 5 ▶ ' α ⬅Z ⬅= α ⬅T − 3 ENTER.

2. Convert to a point and direction vector for the line, then make an extra copy of the direction vector: VAR **P÷PD** ENTER.

3. Rotate the point into level 1, enter the given point (the one not on the given line), and compute the vector between these two points: ⬅STACK **ROT** ⬅[ ] 5 +/− SPC 2 SPC 1 ENTER −.
   Result: **[3 3 −4]**

4. Compute the area of the triangle: MTH **VECTR** **CROSS** **ABS** 2 ÷.

5. Find the height of the triangle, and thus the distance from the given point to the given line: 2 X SWAP **ABS** ÷.
   Result: **5.82482372509**

The program, $\Box t\,o\Box$, first discussed on page 189, can be used to automate the procedure of finding the distance between a point and a line. The following shows how the previous example should be modified in order to use $\Box t\,o\Box$.

Example: Repeat the previous example using the program, $\Box t\,o\Box$.

1. Enter the point in space: ⬅[ ] 5 +/− SPC 2 SPC 1 ENTER.
2. Enter the list of parametric equations: ⬅{ } ' α⬅X ⬅= 3 X α⬅T − 2 ▶ ' α⬅Y ⬅= 2 +/− X α⬅T + 5 ▶ ' α⬅Z ⬅= α⬅T − 3 ENTER.
3. Convert to a point and direction vector for the line: VAR **P÷PD** .
4. Now find the array for the line: ⬅STACK **OVER** + 2 MTH **MATR** **ROW** **ROW÷** .
5. Find the distance from the point to the line with $\Box t\,o\Box$: VAR **DTOL** .

<u>Result:</u> 5.82482372511

Example: Find the distance from the origin to the line $-2x + y - 4 = 0$.

1. Enter the point of the origin [0 0]: ⬅[ ] 0 SPC 0 ENTER.
2. Enter the equation of the line: ' 2 +/− X α⬅X + α⬅Y − 4 ⬅= 0 ENTER.
3. Convert the line to array form: VAR **G÷A** .
4. Execute $\Box t\,o\Box$: **DTOL** . <u>Result:</u> 1.788854382

Given: Two Lines

Two lines, if restricted to the *xy*-plane, are either *concurrent* (i.e. they are the same line), *parallel*, or *intersecting with* respect to one another. If the two *lines* are not restricted in space, they may also be *skew* with respect to one another—that is, they neither intersect nor are parallel.

Before working with a pair of lines, it is important to establish their relationship with one another. The program LIN2? (see page 288) takes equations of the two lines, in array form, and returns an object indicating their relationship on level 2 and a test result on level 1: 0 if the lines don't intersect and 1 if they do. If the lines intersect, then the object returned on level 2 is a vector containing the coordinates of the point of intersection. If the lines do not intersect, then the object on level 2 is a string indicating the relationship (parallel, skew, or concurrent).

Example: Determine the relationship of the following two lines, $4x - 3y + 1 = 0$ and $-2x + 5y + 2 = 0$.

1. Enter the first line and convert to an array: ['] [4] [×] [α] [←] [×] [−] [3] [×] [α] [←] [Y] [+] [1] [←] [=] [0] [ENTER] [VAR] **G→A**.

2. Enter the second line and convert to an array: ['] [2] [+/−] [×] [α] [←] [×] [+] [5] [×] [α] [←] [Y] [+] [2] [←] [=] [0] [ENTER] [VAR] **G→A**.

3. Execute LIN2?: [α] [α] [L] [I] [N] [2] [?] [ENTER] or [VAR] (then [NXT] or [←] [PREV] as needed) **LIN2?**.

   <u>Result:</u>   2: [-.785714 -.714286]
                    1:                        1

   The lines intersect at the point $\left(-\dfrac{11}{14}, -\dfrac{5}{7}\right)$. (As you've seen earlier, you can compute the fractions using the program A→Q.)

Example: Determine the relationship of the following two lines, which are given here parametrically: $\{\, x = 3t + 6,\, y = -2t - 1,\, z = t - 5 \,\}$ and $\{\, x = -t + 3,\, y = 3t - 1,\, z = -4t + 3 \,\}$.

1. Enter the first set of parametric equations: ⇦[{}] ['] [α]⇦[X] ⇦[=]
   [3][X][α]⇦[T][+][6][▶] ['] [α]⇦[Y]⇦[=][2][+/−][X][α]⇦[T][−][1][▶]
   ['] [α]⇦[Z]⇦[=][α]⇦[T][−][5][ENTER].

2. Convert the set to array form: [VAR] **P→PD** ⇦[STACK] **OVER** [+][2]
   [MTH] **MATR ROW ROW→**.

3. Enter the second set of parametric equations: ⇦[{}] ['] [α]⇦[X]⇦
   [=][+/−][α]⇦[T][+][3][▶] ['] [α]⇦[Y]⇦[=][3][X][α]⇦[T][−][1][▶]
   ['] [α]⇦[Z]⇦[=][4][+/−][X][α]⇦[T][+][3][ENTER].

4. Convert the set to array form: [VAR] **P→PD** ⇦[STACK] **OVER** [+][2]
   [MTH] **MATR ROW ROW→**.

5. Execute $\mathsf{LIN2?}$: [VAR]**LIN2?**. Result: 2:   "Skew"
   1:         0

If two lines are parallel, the distance between them is constant, so you can find the distance by identifying a point on one of the lines and using the $\mathsf{Dt\,oL}$ program....

Example: Find the distance between $4x - 8y + 3 = 0$ and $2x - 4y - 6 = 0$.

1. Enter the first line: ['][4][X][α]⇦[X][−][8][X][α]⇦[Y][+][3]⇦[=][0]
   [ENTER].

2. Convert that to an array, then reduce it to one point on the line: [VAR]
   G→A [MTH] **MATR ROW →ROW** [◄][◄]. Result: [1 .875]

3. Enter the other line; convert to an array: ['][2][X][α]⇦[X][−][4][X][α]
   ⇦[Y][−][6]⇦[=][0][ENTER][VAR]**G→A**. Result: [[1 -1]
   [2 -.5]

4. Find the distance from the point to the line (also the distance between the parallel lines): **DTOL**. Result: 1.67705098312

The angle (θ) formed by two intersecting lines can be computed from the definition of the dot product:

$$\theta = \cos^{-1} \frac{\mathbf{A} \bullet \mathbf{B}}{|\mathbf{A}\|\mathbf{B}|}, \text{ where } \mathbf{A} \text{ and } \mathbf{B} \text{ are direction vectors for the two lines.}$$

Example: Compute the angle (in degrees) between the two intersecting lines $4x + 2y - 1 = 0$ and $5x - y + 3 = 0$.

1. Set degree mode (if necessary) by pressing ⬅RAD to turn off the angle annunciator.

2. Enter the first line and find its direction vector: ' 4 ✕ α⬅ ✕ + 2 ✕ α⬅ Y ⊟ 1 ENTER VAR G→A MTH MATR ROW →ROW ◀⊟. Result: `[ -1 2 ]`

3. Enter the second line and compute its direction vector: ' 5 ✕ α ⬅ ✕ ⊟ α⬅ Y + 3 ENTER VAR G→A MTH MATR ROW →ROW ◀⊟. Result: `[ -1 -5 ]`

4. Compute the angle between the two lines: ⬅STACK NXT DUP2 MTH VECTR DOT 3 ⬅STACK ROLLD →MENU ABS SWAP ABS ✕ ÷ ⬅ACOS.

   Result: `142.125` (in degrees, to three decimal places)

   Note that this procedure finds the angle between two vectors sharing initial endpoints and there is only one angle between them, but it is a stand-in for the angle between intersecting lines, which form two angles. Thus the supplementary angle ($\approx$37.875°) is also formed by the intersection of the two lines.

Two lines that either intersect or are parallel determine a unique plane. The easiest approach to finding the equation of the plane given two lines is to find three points —two from one line and one from the other (but not the point of intersection)— and use the procedure described on page 186 when given three non-collinear points. (The following two examples assume that you already know that the two lines either intersect or are parallel.)

Example: Find the equation of the plane determined by the lines { $x = 3t - 2$, $y = -2t + 5$, $z = t - 3$ } and { $x = -t + 14$, $y = 4t - 19$, $z = 5t - 19$ }.

1. Enter the first set of parametric equations: ⬅️{}❜α⬅️X⬅️= 3X α⬅️T➖2▶❜α⬅️Y⬅️=2+/−X α⬅️T➕5▶ ❜α⬅️Z⬅️=α⬅️T➖3 ENTER.

2. Find a position vector (point) for the line; make a copy: VAR **P→PD** ⬅️ ENTER.

3. Enter the other set of parametric equations; convert it to position-direction form: ⬅️{}❜α⬅️X⬅️=+/−α⬅️T➕14▶❜ α⬅️Y⬅️=4X α⬅️T➖19▶❜α⬅️Z⬅️=5X α ⬅️T➖19 ENTER **P→PD**.

4. Compute the set of coefficients [ABC] for the equation of the plane: 3⬅️STACK **ROLLD** ➖ MTH **VECTR** **CROSS**.

   Result: [−56 −64 40]

5. Compute the D coefficient: **DOT** +/−. Result: 328

   Thus an equation for the plane is: $-56x - 64y + 40z - 328 = 0$. Note that this can be reduced by dividing through by 8:

   $$-7x - 8y + 5z + 41 = 0$$

A plane has three traces. A *trace* is the line of intersection of the plane with one of the three coordinate planes (the *xy*-plane, the *yz*-plane, and the *xz*-plane). The following example illustrates how to compute the equation of the plane from its three traces.

Example: Find the equation of the plane that has the following three traces:

$$xy\text{-trace:} \quad 2x - 3y + 12 = 0$$
$$yz\text{-trace:} \quad -3y - 6z + 12 = 0$$
$$xz\text{-trace:} \quad 2x - 6z + 12 = 0$$

1. Write down the *xy*-trace.

2. Inspect the *yz*-trace and find the *z*-term.

3. Insert the *z*-term into the *xy*-trace:

$$\underline{\text{Result:}} \quad 2x - 3y - 6z + 12 = 0$$

The inverse process, finding the traces of a plane given the equation of the plane, is almost as easy.

Example: Find the three traces of the plane, $4x - 5y + z + 1 = 0$.

1. Write down the equation of the plane: $4x - 5y + z + 1 = 0$.

2. To find the *xy*-trace, eliminate the *z*-term (equivalent to letting $z=0$):
$$4x - 5y + 1 = 0.$$

3. To find the *yz*-trace, eliminate the *x*-term (equivalent to letting $x=0$):
$$-5y + z + 1 = 0.$$

4. To find the *xz*-trace, eliminate the *y*-term (equivalent to letting $y=0$):
$$4x + z + 1 = 0.$$

Given: Two Planes

The general equation of a plane is $Ax + By + Cz + D = 0$, often expressed as a vector of its coefficients, $[A B C D]$. The vector represented by $[A B C]$ is the *normal vector* for the plane—the direction vector for a line perpendicular to the plane. Thus, a unique plane is defined by a point in the plane and its normal vector.

Two planes are either parallel or concurrent if the normal vector of one plane is a constant multiple of the normal vector of the other. If the ratio of the D coefficients is equal to this same constant multiple, the planes are concurrent; if the D-ratio is different than the constant multiple, the planes are parallel.

If two planes are not parallel, they intersect in a line which has a direction vector perpendicular to the normal vectors for the two planes. To determine the equation of the line:

1. Determine a point on the line of intersection;
2. Find the cross product of the normal vectors;
3. Convert this to a set of parametric equations for a line.

Example: Find the line of intersection, if any, of these two planes:
$$4x - 5y + z - 2 = 0 \text{ and } x + 2y + 2z = 0$$

1. Determine a point on the line of intersection. Assume that this point has a z-coordinate of 0 and solve the two equations of the plane simultaneously to find the x- and y-coordinates: Enter a vector of the two constant terms: ⬅[ ]2 +/– SPC 0 ENTER. Enter a matrix of the x- and y-coefficients: ➡MATRIX 4 ENTER 5 +/– ENTER ▼1 ENTER 2 ENTER ENTER. Solve the linear system, ÷, then append a 0 to the result as the z-coefficient you assumed: 0 ENTER 3 MTH **MATR COL COL+**. Result: [-.30769 .15385 0]

2. Find the cross product of the normals: ⬅[ ]4 SPC 5 +/– SPC 1 ENTER ⬅[ ]1 SPC 2 SPC 2 ENTER MTH **VECTR CROSS**. Result: [-12 -7 13]

3. Convert the point and vector to a set of parametric equations: VAR **PD÷P**. Result: { 'x=-(12*t)-4/13' 'y=-(7*t)+2/13' 'z=13*t' }

The following example uses the program PL2→L (see page 294), which first determines whether or not two planes (entered as vectors of coefficients) are parallel, and if not, the set of parametric equations for the line of intersection.

Example: Repeat the previous example using PL2→L.

1. Enter the two planes in vector form: ⤆[ ] 4 SPC 5 +/− SPC 1 SPC 2 +/− ENTER; ⤆[ ] 1 SPC 2 SPC 2 SPC 0 ENTER.

2. Execute PL2→L: VAR **PL2**.

   Result: { 'x=-(12*t)-4/13' 'y=-(7*t)+2/13' 'z=13*t' }

The angle formed by the intersection of two planes is easily determined by computing the angle between their normal vectors:

$$\theta = \cos^{-1} \frac{\mathbf{A} \bullet \mathbf{B}}{|\mathbf{A}\|\mathbf{B}|},$$ where $\mathbf{A}$ and $\mathbf{B}$ are normal vectors for the two planes.

Example: Find the angle (in degrees) between the two planes, $x - 2y - 2z + 3 = 0$ and $6x + 3y + 2z - 1 = 0$.

1. Set degree mode (if necessary) by pressing ⤆ RAD to turn off the angle annunciator.

2. Enter the normal vectors of the two planes: ⤆[ ] 1 SPC 2 +/− SPC 2 +/− ENTER; ⤆[ ] 6 SPC 3 SPC 2 ENTER.

3. Compute the angle between the two lines: ⤆ STACK NXT **DUP2** MTH **VECTR** **DOT** 3 ⤆ STACK **ROLLD** ➡ MENU **ABS** SWAP **ABS** × ÷ ⤆ ACOS. Result: 100.981 (in degrees, to 3 places)

Note that this procedure finds the angle between two vectors sharing initial endpoints and there is only one angle between them, but it is a stand-in for the angle between intersecting planes, which form two angles. Thus the supplementary angle (≈79.019°) is also formed by the intersection of the two planes.

Given: A Point and a Plane

Your most common calculation when given a point and a plane is to determine the distance between them, which is given by

$$\text{distance} = \frac{|Aa + Bb + Cc + D|}{\sqrt{A^2 + B^2 + C^2}}$$

where $A, B, C,$ and D are the coefficients of the plane in general form, and $a, b,$ and c are the components of the given point $[\,a\ b\ c\,]$. If you let $\mathbf{N}$ be the normal vector for the plane and $\mathbf{P}$ the position vector for the point, then the distance equa-tion becomes

$$\text{distance} = \frac{|\mathbf{N} \bullet \mathbf{P} + D|}{|\mathbf{N}|}$$

Example: Calculate the distance between the point $[\,4\ \text{-}1\ \text{-}3\,]$ and the plane $-2x + 3y - 2z + 6 = 0$.

1. Enter the normal vector for the plane and make an extra copy: ⬅︎[] 2 +/− SPC 3 SPC 2 +/− ENTER ENTER.

2. Enter the D-constant for the plane and swap it with the copy of the normal vector: 6 ENTER SWAP

3. Enter the point: ⬅︎[] 4 SPC 1 +/− SPC 3 +/− ENTER

4. Compute the distance: MTH VECTR DOT + ABS SWAP ABS ÷. <u>Result</u>: .242535625036

Note: If the resulting distance is zero, then the point lies in the plane.

The next two examples illustrate how to find the equation of a plane given a point in that plane and the equation of a second plane.

Example: Find the equation of the plane parallel to $x - 3y + 2z - 5 = 0$ that contains the point [4 2 -3].

1. Enter the normal vector of the given plane (the same as the normal vector of the parallel plane): (←)[ ][1][SPC][3][+/−][SPC][2][ENTER].

2. Enter the given point: (←)[ ][4][SPC][2][SPC][3][+/−][ENTER].

3. Find the D-constant for the parallel plane: [MTH][VECTR][DOT][+/−]. Result: 8. The equation of the parallel plane is: $x - 3y + 2z + 8 = 0$.

While there is only one plane containing a given point parallel to a given plane, there are an infinite number of planes containing a given point that are perpendicular to a given plane. The cross product is the easiest method of finding one of those perpendicular planes, as the next example illustrates.

Example: Find the equation of a plane perpendicular to $x - 3y + 2z - 5 = 0$ that contains the point [4 2 -3].

1. Enter the given point and make an extra copy: (←)[ ][4][SPC][2][SPC] [3][+/−][ENTER][ENTER].

2. Enter the normal vector of the given plane: (←)[ ][1][SPC][3][+/−] [SPC][2][ENTER].

3. Compute the cross product (this gives a vector perpendicular to the normal vector: [MTH][VECTR][CROSS]. Result: [−5 −11 −14].

4. Compute the D-constant for the perpendicular plane: [DOT][+/−]. Result: 0. This is to be expected because the computed normal is perpendicular to the position vector of the point as well as to the given plane. (The dot product of perpendicular vectors is 0.)

   So the equation of the perpendicular plane is: $-5x - 11y - 14z = 0$

Given: A Line and a Plane

A line and a plane are parallel if the dot product of their direction vector and normal vector is zero and a point on the line does *not* lie on the plane. If the dot product is zero and a point on the line does lie on the plane, the line is contained in the plane; if the dot product is not zero, then the plane and line intersect in a point.

Example: Find the number of points shared by the line { $x = t + 3, y = t - 1, z = -t + 1$ }, and the plane, $3x - 2y + z - 4 = 0$.

1. Enter the normal vector for the plane and make an extra copy: ⏴[ ]
   3 SPC 2 +/− SPC 1 ENTER ENTER.

2. Enter the line and convert it to position-direction form: ⏴{ } ' α
   ⏴X ⏴= α⏴T + 3▶ ' α⏴Y⏴= α⏴T−1▶
   ' α⏴Z⏴=+/−α⏴T +1 ENTER VAR P⇒PD.

3. Find the dot product of the direction vector of the line and the normal vector of the plane: ⏴STACK ROT MTH VECTR DOT.

   Result: 0. The line is either parallel to, or is contained in, the plane.

4. Determine whether the point on the line lies on the plane by finding the negative of the dot product of the point and the normal vector for the plane: ◀ DOT +/−. Result: −12.

   If the point were on the plane, this result would be equal to the *D*-constant in the equation of the plane. That's not the case here, so the line is parallel to the plane; they share no points.

If the dot product of the direction vector of a line and the normal vector of a plane is not zero, then they intersect at a point. To find the point, substitute the parametric equations of the line into the equation for the plane and solve for t. Then substitute the computed value of t back into the parametric equations to find the x-, y-, and z-coordinates of the point of intersection. In vector terms,

$$t = \frac{-D - (\mathbf{N} \bullet \mathbf{P})}{(\mathbf{N} \bullet \mathbf{D})}$$

where D is the D-constant in the equation of the plane, $\mathbf{N}$ is the normal vector of the plane, $\mathbf{P}$ is the position vector of the line and $\mathbf{D}$ is the direction vector of the line. The following example illustrates how to use this equation to compute the coordinates of the intersection point.

Example: Find the point of intersection of the line, $\{\ x = 3t + 1, y = -t - 1, z = -t + 1\ \}$, and the plane, $3x - 2y + z - 4 = 0$.

1. Enter the parametric equations of the line; make an extra copy: ⟵
   {} ' α ⟵X ⟵= 3 × α ⟵T + 1 ▶ ' α ⟵Y ⟵= +/−
   α ⟵T − 1 ▶ ' α ⟵Z ⟵= +/− α ⟵T + 1 ENTER ENTER .

2. Enter the negative of the D-constant and swap it with the copy of the equation of the line: 4 ENTER SWAP .

3. Convert the parametric form to the point direction form: VAR **P⇒PD** .

   <u>Result:</u> 2: [1 -1 1]
                   1: [3 -1 -1]

4. Enter the normal vector of the plane and make an extra copy: ⟵ []
   3 SPC 2 +/− SPC 1 ENTER ENTER .

5. Compute t: 3 ⟵STACK **ROLLD** MTH **VECTR** **DOT** 4 →MENU
   ROLLD →MENU **DOT** − SWAP ÷ . <u>Result:</u> −.2

6. Store the result in t and evaluate the parametric equations: ' α ⟵T
   STO 1 ENTER ⟵ « » EVAL ENTER PRG **LIST** **PROC** **DOLIS** .
   <u>Result:</u> { 'x=.4' 'y=-.8' 'z=1.2' }

7. *Optional.* Convert this to fractions: ⟵SYMBOLIC NXT **→Q** .
   <u>Result:</u> { 'x=2/5' 'y=-(4/5)' 'z=6/5' }

The program LPL→P (see page 288) takes a line in position-direction form and a plane in vector form and, after checking to see if they are parallel, computes the coordinates for the point of intersection. The next example illustrates its use.

Example: Use LPL→P to find the point of intersection, if any, of the line { $x = 2t + 1$, $y = -t - 1$, $z = 3t$ } and the plane $3x + 2y - z - 5 = 0$.

1. Enter the set of parametric equations for the line: [←][{ }]['][α][←][X]
   [←][=][2][×][α][←][T][+][1][▶]['][α][←][Y][←][=][+/−][α][←][T][−][1][▶]
   ['][α][←][Z][←][=][3][×][α][←][T][ENTER].

2. Convert the line to position-direction form: [VAR] **P→PD**.

3. Enter the plane in vector form: [←][[ ]][3][SPC][2][SPC][1][+/−][SPC]
   [5][+/−][ENTER].

4. Execute LPL→P: [α][α][L][P][L][↱][→][P][ENTER] or [VAR] (then [NXT] or
   [←][PREV] as needed) **LPL→**. Result: [9 −5 12]

   If you wish, you can rationalize the results using A→Q.

Example: Find the equation of the plane that contains the line { $x = 2t + 1$, $y = -t - 1$, $z = 3t$ } and is perpendicular to the plane, $3x + 2y - z - 5 = 0$.

1. Enter the normal vector for the plane and make an extra copy: [←][[ ]]
   [3][SPC][2][SPC][1][+/−][ENTER][ENTER].

2. Enter the equation of the line: [←][{ }]['][α][←][X][←][=][2][×][α][←][T]
   [+][1][▶]['][α][←][Y][←][=][+/−][α][←][T][−][1][▶]['][α][←][Z][←][=][3]
   [×][α][←][T][ENTER].

3. Convert the line to position-direction form: [VAR] **P→PD**.

4. Compute the vector of coefficients for the target plane: [←][STACK]
   ROT [MTH] **VECTR CROSS** [3][↱][MENU] **ROLLD** [↱][MENU] **DOT**
   [+/−][4][MTH] **MATR COL COL+**. Result: [−5 11 7 −1]

   So the equation of the perpendicular plane containing the given line is $-5x + 11y + 7z - 1 = 0$.

Introduction to Transformations

So far in this chapter, you've seen how to deal analytically with various combinations of geometric objects. You have used the array object type—vectors and matrices—extensively in the process. In this last part of the chapter, you will learn how to efficiently *transform* groups of points using array methods. Such methods are the foundation of the moving graphics embedded in video games and all kinds of computer modeling.

A *transformation* is a kind of function that maps an input array—representing a geometric object—into a output array. If the transformation is part of a computer graphics program, the program redraws the object, based on the output array; the viewer sees the object undergoing a transformation on the screen.

There are several kinds of transformations possible:

- Translation—moving an object a given distance along a given line.
- Rotation—rotating an object through a given angle around a given axis.
- Reflection—finding the mirror image of an object with respect to a given plane.
- Scaling—changing the size of the object proportionally by a given factor.
- Shearing—changing the size of the object disproportionately.
- Combinations of any or all of the above.

Also, because most visual representations of objects occur in two dimensions (on a display screen or on paper) there are some important transformations to *project* three-dimensional locations into locations that can be plotted in two dimensions:

- Perspective Projection—transformation from three-dimensional space onto a hemispherical surface, where no two lines are parallel, followed by a projection from the hemisphere onto a plane.
- Dimetric Projection—projection which preserves the perpendicularity of the coordinate axes, while equally foreshortening two of the three axes.
- Isometric Projection—projection which preserves the perpendicularity of the coordinate axes, while equally foreshortening all three axes.

For the purposes of mathematical transformation, geometric objects are presented as arrays of points (i.e. arrays of the position vectors of points) with one important addition. An extra element, 1, is appended to the position vector, yielding what is known as the *homogeneous coordinate representation*.

Thus, depending on whether the object is in the xy-plane or in three-dimensional space, each position vector will have either 3 or 4 elements. For example, the line segment connecting the point (4,-1,3) to the point (3,2,-1) is represented as

$\begin{bmatrix} 4 & -1 & 3 & 1 \\ 3 & 2 & -1 & 1 \end{bmatrix}$; likewise, a square in the xy-plane is represented by a 4x3

matrix (four points, three elements each); and a cube is represented by an 8x4 matrix (eight points, four elements each).

The other important component in a transformation is the *transformation matrix*. It is a square matrix with the same number of columns as the object array: There is a 3x3 general transformation matrix for points limited to two dimensions and an analogous 4x4 transformation matrix for points in space:

$$\left[\begin{array}{cc|c} a & b & p \\ c & d & q \\ \hline l & m & s \end{array}\right] \qquad \left[\begin{array}{ccc|c} a & b & e & p \\ c & d & g & q \\ f & h & j & r \\ \hline l & m & n & s \end{array}\right].$$

Every general transformation matrix can be divided into four sections, each of which "controls" aspects of the transformation:

- a, b, c, d, e, f, g, h, and j control the local-scaling, shearing, and rotation aspects;
- l, m and n control the translation aspects;
- p, q and r control the projection aspects;
- s controls the overall scale of the transformation.

The remainder of this chapter illustrates how to use these "controls" to achieve a wide variety of transformations.

Scaling

To isolate the effect of each component of the transformation matrix, you must make sure that all other components have no effect on the transformation. The 3x3 and 4x4 *identity* matrices,

$$\begin{bmatrix} 1 & 0 & 0 \\ 0 & 1 & 0 \\ 0 & 0 & 1 \end{bmatrix} \quad \text{and} \quad \begin{bmatrix} 1 & 0 & 0 & 0 \\ 0 & 1 & 0 & 0 \\ 0 & 0 & 1 & 0 \\ 0 & 0 & 0 & 1 \end{bmatrix}$$

represent the "no effect" matrices. They show the "neutral" values of each of the "controls"—values to use if want them to have no effect on the transformation.

The scaling controls fall along the diagonal of the matrix:

- *a* controls the scale of the *x*-component.
- *d* controls the scale of the *y*-component.
- *j* controls the scale of the *z*-component (if any).
- *s* controls the scale of all components simultaneously.

For example, to triple the scale of the horizontal component only (for an object in space), you would use the following transformation matrix:

$$\begin{bmatrix} 3 & 0 & 0 & 0 \\ 0 & 1 & 0 & 0 \\ 0 & 0 & 1 & 0 \\ 0 & 0 & 0 & 1 \end{bmatrix}$$

Visually, you would see the object "stretching" horizontally.

The examples in the rest of this chapter use the program TVIEW (see page 316), which draws a simple object, given an array of its points, thus allowing you to view the results of your transformations on that object.

Example: Plot a simple unit square, then "stretch" it horizontally by a factor of three and view the results.

1. Enter the array of points for a unit square (in the *xy*-plane):  [→][MATRIX] [0][ENTER][0][ENTER][1][ENTER][▼][1][ENTER][0][ENTER][1][ENTER][1] [ENTER][1][ENTER][1][ENTER][0][ENTER][1][ENTER][1][ENTER][ENTER].

   Result:
   ```
   [[ 0 0 1 ]
    [ 1 0 1 ]
    [ 1 1 1 ]
    [ 0 1 1 ]]
   ```

2. Plot the square with **TVIEW**: [α][α][T][V][I][E][W][ENTER] or [VAR] (then [NXT] or [←][PREV] as needed) **TVIEW**.

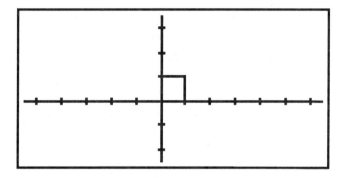

3. At the stack, enter the transformation matrix: [CANCEL][→][MATRIX] [3][ENTER][0][ENTER][0][ENTER][▼][0][ENTER][1][ENTER][0][ENTER][0] [ENTER][0][ENTER][1][ENTER][ENTER]. Result:
   ```
   [[ 3 0 0 ]
    [ 0 1 0 ]
    [ 0 0 1 ]]
   ```

4. Multiply the object array by the transformation: [×] **TVIEW**.

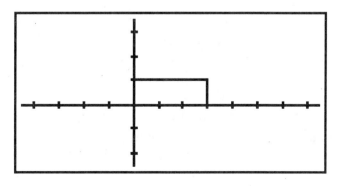

Of course, the y-component can be scaled as well by changing the value of the d element in the transformation matrix.

Example: Stretch the rectangular result of the previous example vertically by a factor of 2 and display the results.

1. Return to the stack and enter the appropriate 3x3 transformation matrix: (CANCEL) (→)(MATRIX) (1)(ENTER) (0) (ENTER)(0) (ENTER)(▼) (0) (ENTER) (2) (ENTER)(0)(ENTER) (0)(ENTER) (0)(ENTER) (1)(ENTER)(ENTER).

   Result: [[1 0 0]
            [0 2 0]
            [0 0 1]]

2. Multiply the object array by the transformation matrix and view the results with TVIEW: (×)**TVIEW**.

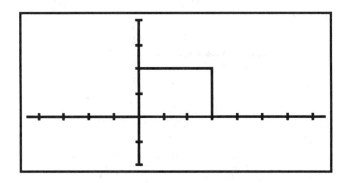

Notice that you could have performed both of the previous transformations in one step by using a transformation matrix initially that is the combination of the two local-scaling steps:

$$\begin{bmatrix} 3 & 0 & 0 \\ 0 & 1 & 0 \\ 0 & 0 & 1 \end{bmatrix}\begin{bmatrix} 1 & 0 & 0 \\ 0 & 2 & 0 \\ 0 & 0 & 1 \end{bmatrix} = \begin{bmatrix} 3 & 0 & 0 \\ 0 & 2 & 0 \\ 0 & 0 & 1 \end{bmatrix}$$

This is a characteristic of combination transformations—they are the product of the individual component transformations.

6. *ANALYTIC GEOMETRY*

The lower right-hand element of the transformation matrix (the *s* element) controls the overall scaling of the object. It works differently than the local scaling factors. It is located in the last column, which should be all 1 's unless a projection is intended. To restore the expected state of the last column, all elements of the matrix are divided by the *s*-factor, thus effectively making the lower-right-hand element a 1 again.

Example: Explore the effects of transforming the current object using a global-scaling factor of 2.

1. Return to the stack and enter the appropriate 3x3 transformation matrix: (CANCEL) (→)(MATRIX) (1)(ENTER) (0) (ENTER)(0) (ENTER)(▼) (0) (ENTER)(1)(ENTER)(0)(ENTER)(0)(ENTER)(0)(ENTER)(2)(ENTER)(ENTER).

Result: [[1 0 0]
         [0 1 0]
         [0 0 2]]

2. Multiply the object array by the transformation matrix and view the results with TVIEW: (×)**TVIEW**.

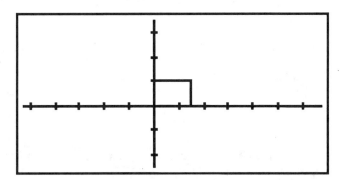

Note that, because of the division process involved, a global-scaling factor *greater* than one *shrinks* the object; and a global scaling factor *less* than one *enlarges* the object.

Shearing

The pure scaling you've seen already—using the diagonal elements of the transformation matrix—represent the "self-effects" of scaling of individual coordinates only: *independent* scaling.

By contrast, *shearing* is the effect that the scaling of one coordinate has on the value of other coordinates. Shearing is *dependent* scaling, using the off-diagonal elements b, c, e, f, g, and h in the transformation matrix:

- b represents the effect of x-scaling on the y-coordinate.
- c represents the effect of y-scaling on the x-coordinate.
- e represents the effect of x-scaling on the z-coordinate (if any).
- f represents the effect of z-scaling (if any z-coordinate) on the x-coordinate.
- g represents the effect of y-scaling on the z-coordinate (if any).
- h represents the effect of z-scaling (if any z-coordinate) on the y-coordinate.

Example: Shear the y-coordinates of the current object by a factor of 1.2 of the x-coordinates.

1. At the stack, enter the transformation matrix: (CANCEL) (→)(MATRIX)
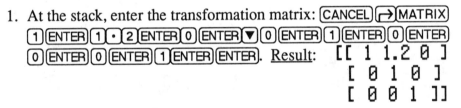

2. Multiply the object array by the transformation matrix and view the results with TVIEW: (×)(TVIEW).

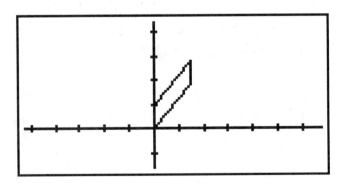

Example: Shear the *x*-coordinates of the current object by a factor of 0.75 of the *y*-coordinates.

1. At the stack, enter the transformation matrix: (CANCEL) (→)(MATRIX) (1)(ENTER)(0)(ENTER)(0)(ENTER)(▼)(.)(7)(5)(ENTER)(1)(ENTER)(0)(ENTER) (0)(ENTER)(0)(ENTER)(1)(ENTER)(ENTER).

    Result: [[1 0 0]
            [.75 1 0]
            [0 0 1]]

2. Multiply the object array by the transformation matrix and view the results with TVIEW: (×)**TVIEW**.

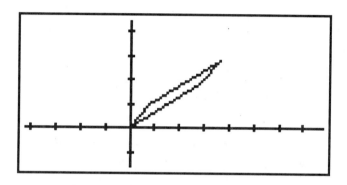

Translation

Translation is a transformation that moves a point along a line. It changes the coordinate system in which the object exists without changing any of the dimensions or relative positions of the points of the object. For example, translating an object to the point (4,5) in the *xy*-plane means that the origin (0,0) is now at the point (4,5) and all points in the object that had been situated with respect to the origin are now analogously situated with respect to the point (4,5).

In the transformation matrix, the *l*-, *m*-, and *n*-elements control the *x*-, *y*-, and *z*-axis translations, respectively. For the above example—a translation to the point (4,5) in the *xy*-plane—the transformation matrix would be:

$$\begin{bmatrix} 1 & 0 & 0 \\ 0 & 1 & 0 \\ 4 & 5 & 1 \end{bmatrix}$$

Example: Translate the current object to a coordinate system centered at the point (2,-1).

1. At the stack, enter the transformation matrix: CANCEL →MATRIX
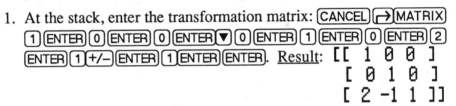
1 ENTER 0 ENTER 0 ENTER ▼ 0 ENTER 1 ENTER 0 ENTER 2 ENTER 1 +/− ENTER 1 ENTER ENTER. Result: `[[ 1  0  0 ]`
`[ 0  1  0 ]`
`[ 2 -1  1 ]]`

2. Multiply the object array by the transformation matrix and view the results with TVIEW: ✕ TVIEW.

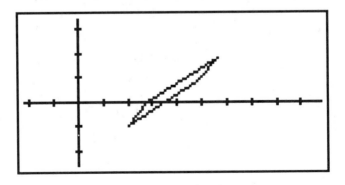

Rotation

A *rotation* is a combination of scaling and shearing that leaves the final dimensions of the object unchanged. Each rotation occurs around an *axis of rotation*. In two-dimensions, it appears that the rotation is around a point—often the origin—but actually, it is occurring around the z-axis, which extends "upward" from the flat xy-plane. If θ is the angle through which you wish to rotate the object counterclockwise around the origin, the appropriate 3x3 transformation matrix is

$$\begin{bmatrix} \cos\theta & \sin\theta & 0 \\ -\sin\theta & \cos\theta & 0 \\ 0 & 0 & 1 \end{bmatrix}$$

Example: Rotate the current object 130° counterclockwise around the origin (z-axis).

1. At the stack, be sure you're in degree mode, and enter the transformation matrix: (CANCEL)(←)(RAD) (if necessary) (→)(MATRIX)(1)(3)(0) (COS)(ENTER)(1)(3)(0)(SIN)(ENTER)(0)(ENTER)(▼)(1)(3)(0)(SIN)(+/−)(ENTER) (1)(3)(0)(COS)(ENTER)(0)(ENTER)(0)(ENTER)(0)(ENTER)(1)(ENTER)(ENTER).

   <u>Result:</u> [[.6428 .7660 0]
   [-.7660 .6428 0]
   [0 0 1]]

2. Multiply the object array by the transformation matrix and view the results with TVIEW: (×)**TVIEW**.

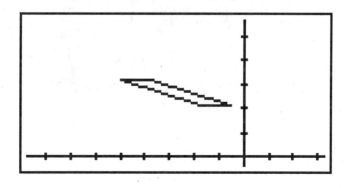

When performing a rotation in the *xy*-plane around an arbitrary vertical axis (which appears to be a point), the rotation is a three-step procedure:

- Translate from the given "point" of rotation back to the origin;
- Rotate the given number of degrees about the origin;
- Translate back to the coord-inate system of the given point.

Thus, if the "point" of rotation is (l,m) and the angle of rotation is θ, the three transformation matrices, in order, are:

$$\begin{bmatrix} 1 & 0 & 0 \\ 0 & 1 & 0 \\ -l & -m & 1 \end{bmatrix} \begin{bmatrix} \cos\theta & \sin\theta & 0 \\ -\sin\theta & \cos\theta & 0 \\ 0 & 0 & 1 \end{bmatrix} \begin{bmatrix} 1 & 0 & 0 \\ 0 & 1 & 0 \\ l & m & 1 \end{bmatrix}, \text{ which, after multiplication,}$$

$$\text{simplify to} \begin{bmatrix} \cos\theta & \sin\theta & 0 \\ -\sin\theta & \cos\theta & 0 \\ -l(\cos\theta - 1) + m\sin\theta & -l\sin\theta - m(\cos\theta - 1) & 1 \end{bmatrix}$$

A program, ROT2D (see page 299), creates the proper 3x3 transformation matrix, given the point (in vector form) on level 2 and the angle of rotation on level 1.

Example: Rotate the current object around the point, (-1,-1) by 50°.

1. At the stack, enter the point of rotation in vector form: [CANCEL] [←] [[ ]] [1] [+/−] [SPC] [1] [+/−] [ENTER].

2. In degree mode, enter the angle: [←] [RAD] (if needed) [5] [0] [ENTER].

3. Use ROT2D: [VAR] [ROT2D]: `[[ .64279 .76604 0 ]`
                           `[ -.76604 .64279 0 ]`
                           `[ -1.12326 .40883 1 ]]`

4. Multiply and view the results: [×] [VIEW].

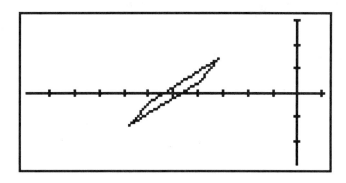

The rotation of objects in three dimensions presents even more complications. A rotation in three dimensions implies that rotations may occur around any of the three coordinate axes—in the most general case, around any line in space.

In a rotation about any one of the coordinate axes, the coordinates of the object with respect to that axis do not change (thus, in a rotation about the z-axis in the xy-plane, where $z = 0$, the implied z-coordinate remains at zero before and after the rotation). Therefore, the rotation matrices for rotating about each of the coordinate axes individually are:

$$
x-axis: \quad
\begin{bmatrix}
1 & 0 & 0 & 0 \\
0 & \cos\theta & \sin\theta & 0 \\
0 & -\sin\theta & \cos\theta & 0 \\
0 & 0 & 0 & 1
\end{bmatrix}
\quad
y-axis: \quad
\begin{bmatrix}
\cos\phi & 0 & -\sin\phi & 0 \\
0 & 1 & 0 & 0 \\
\sin\phi & 0 & \cos\phi & 0 \\
0 & 0 & 0 & 1
\end{bmatrix}
\quad
z-axis: \quad
\begin{bmatrix}
\cos\psi & \sin\psi & 0 & 0 \\
-\sin\psi & \cos\psi & 0 & 0 \\
0 & 0 & 1 & 0 \\
0 & 0 & 0 & 1
\end{bmatrix}
$$

So if you need to rotate about more than one axis—θ about the x-axis, ϕ about the y-axis, and ψ about the z-axis, for example—you simply multiply the necessary transformation matrices together. But note that the order in which you do the successive rotations definitely matters: remember that matrix multiplication is *non-commutative*.

For example, compare two transformations:

Case 1. Rotate about the x-axis, then by an equal angle about the y-axis ($\phi = \theta$):

$$\begin{bmatrix} 1 & 0 & 0 & 0 \\ 0 & \cos\theta & \sin\theta & 0 \\ 0 & -\sin\theta & \cos\theta & 0 \\ 0 & 0 & 0 & 1 \end{bmatrix} \begin{bmatrix} \cos\phi & 0 & -\sin\phi & 0 \\ 0 & 1 & 0 & 0 \\ \sin\phi & 0 & \cos\phi & 0 \\ 0 & 0 & 0 & 1 \end{bmatrix}$$

$$= \begin{bmatrix} \cos\theta & 0 & -\sin\theta & 0 \\ \sin^2\theta & \cos\theta & \cos\theta\sin\theta & 0 \\ \cos\theta\sin\theta & -\sin\theta & \cos^2\theta & 0 \\ 0 & 0 & 0 & 1 \end{bmatrix}$$

Case 2. Rotate about the y-axis, then by an equal angle about the x-axis ($\theta = \phi$):

$$\begin{bmatrix} \cos\phi & 0 & -\sin\phi & 0 \\ 0 & 1 & 0 & 0 \\ \sin\phi & 0 & \cos\phi & 0 \\ 0 & 0 & 0 & 1 \end{bmatrix} \begin{bmatrix} 1 & 0 & 0 & 0 \\ 0 & \cos\theta & \sin\theta & 0 \\ 0 & -\sin\theta & \cos\theta & 0 \\ 0 & 0 & 0 & 1 \end{bmatrix}$$

$$= \begin{bmatrix} \cos\theta & \sin^2\theta & -\cos\theta\sin\theta & 0 \\ 0 & \cos\theta & \sin\theta & 0 \\ \sin\theta & -\cos\theta\sin\theta & \cos^2\theta & 0 \\ 0 & 0 & 0 & 1 \end{bmatrix}$$

Note that these overall transformation matrices are *not* equivalent. Remember this when performing rotations around more than one axis.

The two-dimensional general rotation matrix is a bit complicated; the three-dimensional case is definitely so: For an axis (line) of rotation with a direction vector $[x\,y\,z]$ and a position vector $[l\,m\,n]$, and an angle of rotation θ in the counterclockwise direction, the 4x4 general rotation transformation matrix is:

$$\begin{bmatrix} x^2+(1-x^2)\cos\theta & xy(1-\cos\theta)+z\sin\theta & xz(1-\cos\theta)-y\sin\theta & 0 \\ xy(1-\cos\theta)-z\sin\theta & y^2+(1-y^2)\cos\theta & yz(1-\cos\theta)+x\sin\theta & 0 \\ xz(1-\cos\theta)+y\sin\theta & yz(1-\cos\theta)-x\sin\theta & z^2+(1-z^2)\cos\theta & 0 \\ L & M & N & 1 \end{bmatrix}$$

where
$$L =$$
$$l-l\left[x^2+(1-x^2)\cos\theta\right]-m\left[xy(1-\cos\theta)-z\sin\theta\right]-n\left[xz(1-\cos\theta)+y\sin\theta\right]$$
$$M =$$
$$m-l\left[xy(1-\cos\theta)+z\sin\theta\right]-m\left[y^2+(1-y^2)\cos\theta\right]-n\left[yz(1-\cos\theta)-x\sin\theta\right]$$
$$N =$$
$$n-l\left[xz(1-\cos\theta)-y\sin\theta\right]-m\left[yz(1-\cos\theta)+x\sin\theta\right]-n\left[z^2+(1-z^2)\cos\theta\right]$$

Clearly a program is called for: ROT3D (see page 300) takes the position and direction vectors of the axis of rotation from levels 3 and 2, and the angle of rotation from level 1.

Example: Find the proper transformation matrix to rotate an object around the line given by { $x = 2t + 1$, $y = -t - 1$, $z = 3t$ } by 74°.

1. At the stack, enter the line in parametric form: CANCEL ← { } ' α ← X ← = 2 X α ← T + 1 ▶ ' α ← Y ← = +/– α ← T – 1 ▶ ' α ← Z ← = 3 X α ← T ENTER.

2. Convert the line to position-direction form: VAR **P÷PD**.

2. In degree mode, enter the angle: ← RAD (if necessary) 7 4 ENTER.

3. Use ROT3D to find the transformation matrix: **ROT3D**.

Result: [[3.17309 1.43506 5.30744 0]
  [-4.33251 1 -.25056 0]
  [3.38491 -4.09561 6.79490 0]
  [-6.50560 -1.43506 -5.55800 1]]

Reflection

A rotation moves an object with respect to a line. A *reflection*, on the other hand, moves an object with respect to a *plane*. A reflection may appear to be with respect to a line within the xy-plane, but this is because the line is the trace of the plane of reflection—the only part of the reflection plane within the xy-plane. Thus, what appears to be a reflection with respect to, say, the x-axis (i.e. the line $y = 0$) is actually a reflection with respect to the plane $y = 0$ ($0x + y + 0z + 0 = 0$).

The transformation matrices for reflection across each of the coordinate axes (or the planes with which they are associated) are fairly straightforward:

$$
\text{across } x\text{-axis:} \quad
\begin{bmatrix} 1 & 0 & 0 & 0 \\ 0 & -1 & 0 & 0 \\ 0 & 0 & 1 & 0 \\ 0 & 0 & 0 & 1 \end{bmatrix}
\quad
\text{across } y\text{-axis:} \quad
\begin{bmatrix} -1 & 0 & 0 & 0 \\ 0 & 1 & 0 & 0 \\ 0 & 0 & 1 & 0 \\ 0 & 0 & 0 & 1 \end{bmatrix}
\quad
\text{across } z\text{-axis:} \quad
\begin{bmatrix} 1 & 0 & 0 & 0 \\ 0 & 1 & 0 & 0 \\ 0 & 0 & -1 & 0 \\ 0 & 0 & 0 & 1 \end{bmatrix}
$$

(The 3x3 transformation matrices are the same as those for the x-axis and y-axis shown above, except they have the third row and column removed.)

Example: Reflect the current object across both the x- and y-axes.

1. Enter the 3x3 transformation matrix for reflecting across the x-axis: ([DROP] the 3D-rotation matrix from the previous example, if necessary) [→][MATRIX] [1][ENTER] [0][ENTER] [0][ENTER] [▼][0][ENTER] [1][+/−][ENTER] [0][ENTER] [0][ENTER] [0][ENTER] [1][ENTER] [ENTER].

2. Enter the 3x3 transformation matrix for reflecting across the y-axis: [→][MATRIX] [1][+/−][ENTER] [0][ENTER] [0][ENTER] [▼][0][ENTER] [1][ENTER] [0][ENTER] [0][ENTER] [0][ENTER] [1][ENTER] [ENTER].

3. Because this is just two consecutive reflections, you can multiply the two transformation matrices together before applying it to the object matrix: [×]. <u>Result:</u>
```
[[ -1  0  0 ]
 [  0 -1  0 ]
 [  0  0  1 ]]
```

4. Multiply and view the results: $\boxed{\times}$ $\boxed{\text{TVIEW}}$.

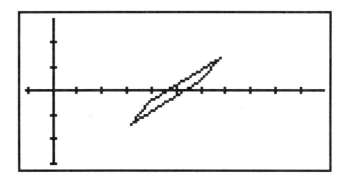

This example was simple, but to reflect across an *arbitrary* plane (or, in two dimensions, an arbitrary line) the computation is complicated. You must rotate the given plane of reflection to match one of the coordinate planes, perform the reflection and rotate the result back. A better approach uses geometric relationships: The plane of reflection is a perpendicular bisector of a line segment from a point to its reflection. The program RFLCT (see page 299) uses this to compute the reflection directly. It takes the object array from level 2 and a vector of the general form of the plane (or, in two dimensions, the line) of reflection from level 1.

Example: Reflect the current object across the line, $x - 2y + 1 = 0$.

1. Return to the stack (where the current object array should be shown in level 1) and enter the vector form of the line of reflection: $\boxed{\text{CANCEL}}$ $\boxed{\leftarrow}$$\boxed{[\ ]}$$\boxed{1}$$\boxed{\text{SPC}}$$\boxed{2}$$\boxed{+/-}$$\boxed{\text{SPC}}$$\boxed{1}$$\boxed{\text{ENTER}}$.

2. Use the RFLCT program: $\boxed{\alpha}$$\boxed{\alpha}$$\boxed{R}$$\boxed{F}$$\boxed{L}$$\boxed{C}$$\boxed{T}$$\boxed{\text{ENTER}}$ or $\boxed{\text{VAR}}$ (then $\boxed{\text{NXT}}$ or $\boxed{\leftarrow}$$\boxed{\text{PREV}}$ as needed) $\boxed{\text{RFLCT}}$.

3. View the results with TVIEW: $\boxed{\text{TVIEW}}$.

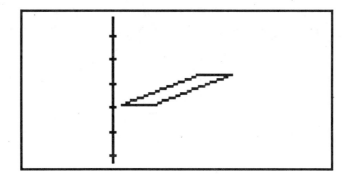

Projection

So far, you have not seen any examples that plot points in three dimensions. Before you can do so, you be able to *project* them into points in two dimensions, so that they may be displayed on the HP 48's two dimensional screen. The kinds of transformations you have seen thus far in this chapter have been *affine* transformations, where the p-, q-, and r-elements in the general 4x4 transformation matrix are zero. But when one or more of these elements is nonzero, then the transformation becomes a *projection*.

Projections depend on two things: the *viewing plane* onto which you're projecting and the center of the projection or *eyepoint*. When you view the HP 48 display, you are viewing the xy-plane (the plane, $z = 0$). Parallel lines stay parallel on the display—they don't meet together in a point. The eyepoint is infinitely far from the viewing plane. Projections that maintain the eyepoint at infinity are called *axonometric projections*, because they keep the coordinate axes at right angles to each other. There are three types of axonometric projections:

- *Orthographic*. These produce the views commonly used in mechanical drawing—Top View, Side View, Front View. They are projections onto one of the three coordinate zero planes ($x = 0$, $y = 0$, or $z = 0$).

- *Dimetric*. These foreshorten two of the three coordinate axes by the same factor, while leaving the axes at right angles (orthographic) to each other. A dimetric projection consists of two successive rotations (once around each of the axes being foreshortened) using angles computed to maintain the orthography of the axes.

- *Isometric*. These foreshorten all three axes by the same factor while maintaining their orthography. An isometric projection is similar to the dimetric projection except that the computed angles of rotation are constant no matter the degree of foreshortening.

The transformation matrix for an orthographic projection onto the xy-plane is:

$$\begin{bmatrix} 1 & 0 & 0 & 0 \\ 0 & 1 & 0 & 0 \\ 0 & 0 & 0 & 0 \\ 0 & 0 & 0 & 1 \end{bmatrix}$$

Note that column corresponding to the projection plane (z) are filled with zeroes, since, in the xy plane, $z = 0$. The analogous approach can be used to construct the matrices for projections onto the $x = 0$ and $y = 0$ planes—leaving the column corresponding to the projection plane filled with zeroes.

Example: Enter this set of points and view it in an orthographic projection onto the xy-plane:

$$\begin{bmatrix} 0 & 0 & 0 & 1 \\ 0 & 0 & 2 & 1 \\ 2 & 0 & 2 & 1 \\ 2 & 0 & 0 & 1 \\ 2 & 2 & 0 & 1 \\ 2 & 2 & 2 & 1 \\ 0 & 2 & 2 & 1 \\ 0 & 2 & 0 & 1 \\ 0 & 0 & 0 & 1 \\ 2 & 0 & 0 & 1 \\ 2 & 2 & 0 & 1 \\ 0 & 2 & 0 & 1 \\ 0 & 2 & 2 & 1 \\ 0 & 0 & 2 & 1 \\ 2 & 0 & 2 & 1 \\ 2 & 2 & 2 & 1 \end{bmatrix}$$

1. Enter the array of points: (→)(MATRIX) (0) (SPC) (0) (SPC) (0) (SPC) (1)
(ENTER)(▼) (0) (SPC) (0) (SPC) (2) (SPC) (1) (ENTER) (2) (SPC) (0) (SPC) (2) (SPC)
(1) (ENTER) (2) (SPC) (0) (SPC) (0) (SPC) (1) (ENTER) (2) (SPC) (2) (SPC) (0) (SPC)
(1) (ENTER) (2) (SPC) (2) (SPC) (2) (SPC) (1) (ENTER) (0) (SPC) (2) (SPC) (2) (SPC)
(1) (ENTER) (0) (SPC) (2) (SPC) (0) (SPC) (1) (ENTER) (0) (SPC) (0) (SPC) (0) (SPC)
(1) (ENTER) (2) (SPC) (0) (SPC) (0) (SPC) (1) (ENTER) (2) (SPC) (2) (SPC) (0) (SPC)
(1) (ENTER) (0) (SPC) (2) (SPC) (0) (SPC) (1) (ENTER) (0) (SPC) (2) (SPC) (2) (SPC)
(1) (ENTER) (0) (SPC) (0) (SPC) (2) (SPC) (1) (ENTER) (2) (SPC) (0) (SPC) (2) (SPC)
(1) (ENTER) (2) (SPC) (2) (SPC) (2) (SPC) (1) (ENTER) (ENTER).

2. Execute TVIEW. It performs an orthographic projection onto the xy-plane simply by ignoring the z-element of each point: (VAR) **TVIEW**.

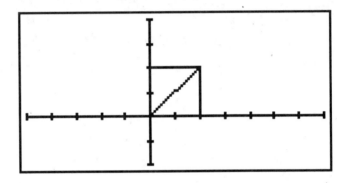

The object is a cube two units on a side with one of its major diagonals drawn, shown here "flattened out" into the xy-plane.

Related to the orthographic projections are those onto planes parallel to the coordinate planes, such as $x = l$, $y = m$, and $z = n$. The transformation matrices for these projections are:

$$\begin{bmatrix} 1 & 0 & 0 & 0 \\ 0 & 0 & 0 & 0 \\ 0 & 0 & 1 & 0 \\ l & 0 & 0 & 1 \end{bmatrix} \quad \begin{bmatrix} 1 & 0 & 0 & 0 \\ 0 & 1 & 0 & 0 \\ 0 & 0 & 0 & 0 \\ 0 & m & 0 & 1 \end{bmatrix} \quad \begin{bmatrix} 1 & 0 & 0 & 0 \\ 0 & 1 & 0 & 0 \\ 0 & 0 & 0 & 0 \\ 0 & 0 & n & 1 \end{bmatrix}$$

The preceding projections share the disadvantage that they lose the z-coordinate information during the projection (although TYIEW avoids this by returning the object array *before* projection while displaying the results *after* projection). A better approach is to use the z-axis information during the projection so that the results give some visual clue about the "depth" of the object.

The *dimetric* projection is of a rotation about the y-axis by an angle ϕ, followed by a rotation about the x-axis by an angle θ. The angles ϕ and θ are computed so that the x- and y- axes are foreshortened by an equal factor, f, while maintaining the orthography of the coordinate axes and projecting the results into the xy-plane.

Obviously, the key is choosing the correct angles. They must satisfy these two equations:

$$\cos^2\phi + \sin^2\phi\sin^2\theta = \cos^2\theta$$
$$\sin^2\phi + \cos^2\phi\sin^2\theta = f^2$$

Once you have computed these angles for a given factor f, you can compute the transformation matrix for the dimetric projection, which is nothing more than the combined matrix from the two rotations about the axes:

$$\begin{bmatrix} \cos\phi & \sin\phi\sin\theta & -\sin\phi\cos\theta & 0 \\ 0 & \cos\theta & \sin\theta & 0 \\ \sin\phi & -\cos\phi\sin\theta & \cos\phi\cos\theta & 0 \\ 0 & 0 & 0 & 1 \end{bmatrix}$$

The program DMTRC (see page 280) computes the appropriate transformation matrix, given the factor, f ($0 \leq f \leq 1$) by which you wish to foreshorten the axes.

Example: Project the current object using a dimetric projection with a factor of 0.5.

1. Return to the stack, make an extra copy of the object array, and store it as CUBE: (CANCEL) (ENTER) ' α α C U B E (ENTER) (STO).

2. Enter the projection factor: (·) (5) (ENTER).

3. Find the transformation matrix for the dimetric projection: α α D (M)(T)(R)(C) (ENTER) or (VAR) (then (NXT) or ← (PREV) as needed) **DMTR**.

Result: [[.92582 .13363 -.35355 0]
[0 .93541 .35355 0]
[.37796 -.32733 .86603 0]
[0 0 0 1]].

4. Multiply the object array by the transformation matrix and display the results using TVIEW: (×) **TVIEW**

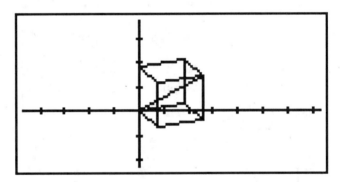

The *isometric* projection is similar to the dimetric projection except that it needs no factor: there is only one set of angles, θ and ϕ, that will equally foreshorten all three axes without disturbing coordinate orthogonality. The required values are $\theta = 35.26439°$ and $\phi = 45°$.

The program I SMTRC (see page 285) takes nothing from the stack and returns the proper transformation matrix for an isometric projection.

Example: Project the CUBE using an isometric projection.

1. Return to the stack, drop the previous result array, and put CUBE onto the stack as the object array: (CANCEL)(◄)(VAR) **CUBE** .

2. Compute the transformation matrix for the isometric projection: (α) (α)(I)(S)(M)(T)(R)(C)(ENTER) or (VAR) (then (NXT) or (◄)(PREV) as needed) **ISMTR**.

   <u>Result:</u> [[.70711 .40825 -.57735 0]
   [0 .81650 .57735 0]
   [.70711 -.40825 .57735 0]
   [0 0 0 1]].

3. Multiply the object array by the transformation matrix and display the results using TVIEW: (×)**TVIEW**.

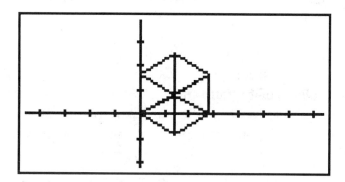

The final set of projections to illustrate are the *perspective projections*. Perspective projections are combinations of a perspective transformation with a projection into a plane. Widely used to present data in visually useful ways, perspective projections are often combined with other transformations—rotations, translations, or scaling—before the perspective transformation (and sometimes after).

The simplest perspective projection projects onto the *xy*-plane from an eyepoint at [0 0 -*k*] where *k* is a finite number. (By contrast, the theoretical eyepoint for the previous axonometric projections was [0 0 -∞].) In this projection, lines that were originally parallel to the *z*-axis will now appear to pass through the *vanishing point* [0 0 *k*].

This projection, known as a *single-point perspective transformation*, is accomplished using the following transformation matrix:

$$\begin{bmatrix} 1 & 0 & 0 & 0 \\ 0 & 1 & 0 & 0 \\ 0 & 0 & 0 & \frac{1}{k} \\ 0 & 0 & 0 & 1 \end{bmatrix}$$

Note the two differences between this matrix and the 4x4 identity matrix: the *r*-element has a nonzero value, and the third column is all 0's (for projection onto the *z* = 0 plane).

However, any non-zero values in the final column (except for the final row) of the transformation give an undesirable scaling effect, producing values other than 1 in the fourth column of the transformed object array. To counteract that effect, the result of a perspective transformation must be *normalized* by dividing the *x*- *y*- and *z*-coordinates of each point by the value of the fourth element in its row.

The short program NRMLZ (see page 289) performs this normalizing procedure on the array in level 1. The resulting array will have its final column filled with ones.

Example: Project the current object onto the *xy*-plane using a single-point perspective projection. Let $k = 10$.

1. Return to the stack, drop the previous result array, and put CUBE onto the stack as the object array: (CANCEL)(←)(VAR)**CUBE** .

2. Enter the transformation matrix: (→)(MATRIX)(1)(ENTER)(0)(ENTER)(0) (ENTER)(0)(ENTER)(▼)(0)(ENTER)(1)(ENTER)(0)(ENTER)(0)(ENTER)(0) (ENTER)(0)(ENTER)(0)(ENTER)(·)(1)(ENTER)(0)(ENTER)(0)(ENTER)(0) (ENTER)(1)(ENTER)(ENTER). <u>Result</u>:
```
[[ 1 0 0 0 ]
 [ 0 1 0 0 ]
 [ 0 0 0 .1 ]
 [ 0 0 0 1 ]]
```

3. Multiply; normalize: (×)**NRML**:
```
[[ 0    0    0 1 ]
 [ 0    0    0 1 ]
 [ 1.67 0    0 1 ]
 [ 2    0    0 1 ]
 [ 2    2    0 1 ]
 [ 1.67 1.67 0 1 ]
 [ 0    1.67 0 1 ]
 [ 0    2    0 1 ]
 [ 0    0    0 1 ]
 [ 2    0    0 1 ]
 [ 2    2    0 1 ]
 [ 0    2    0 1 ]
 [ 0    1.67 0 1 ]
 [ 0    0    0 1 ]
 [ 1.67 0    0 1 ]
 [ 1.67 1.67 0 1 ]]
```

4. Display the results using TVIEW: **TVIEW**.

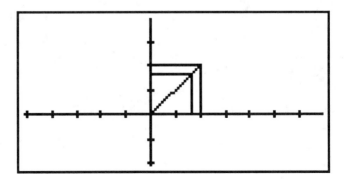

The single-point perspective projection is the one most commonly used by artists, but they effectively translate the object so that the eye-point and the vanishing point are centered on the object. Without such a centering translation, the vanishing point is along the z-axis (i.e. at the origin), no matter where the object is.

Example: Repeat the previous example, but "center" the eyepoint on the object before performing the projection. That is, move the object so that the origin is at its center in the xy-plane—a (-1,-1,0) translation.

1. Return to the stack, drop the previous result array, and put CUBE onto the stack as the object array: CANCEL ← VAR **CUBE**.

2. Prepare the single-point projection, including the translation elements in the last row: → MATRIX 1 ENTER 0 ENTER 0 ENTER 0 ENTER ▼ 0 ENTER 1 ENTER 0 ENTER 0 ENTER 0 ENTER 0 ENTER 0 ENTER . 1 ENTER 1 +/− ENTER 1 +/− ENTER 0 ENTER 1 ENTER ENTER.   Result:

$$\begin{bmatrix} 1 & 0 & 0 & 0 \\ 0 & 1 & 0 & 0 \\ 0 & 0 & 0 & .1 \\ -1 & -1 & 0 & 1 \end{bmatrix}$$

3. Multiply by the object array and normalize: × **NRML**.

Result:

$$\begin{bmatrix} -1 & -1 & 0 & 1 \\ -.83 & -.83 & 0 & 1 \\ .83 & -.83 & 0 & 1 \\ 1 & -1 & 0 & 1 \\ 1 & 1 & 0 & 1 \\ .83 & .83 & 0 & 1 \\ -.83 & .83 & 0 & 1 \\ -1 & 1 & 0 & 1 \\ -1 & -1 & 0 & 1 \\ 1 & -1 & 0 & 1 \\ 1 & 1 & 0 & 1 \\ -1 & 1 & 0 & 1 \\ -.83 & .83 & 0 & 1 \\ -.83 & -.83 & 0 & 1 \\ .83 & -.83 & 0 & 1 \\ .83 & .83 & 0 & 1 \end{bmatrix}$$

4. Display the results using TVIEW: **TVIEW**.

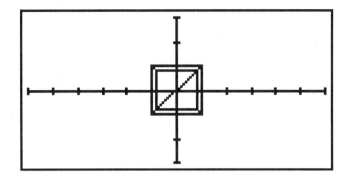

The single-point perspective is equivalent to viewing an object with only one eye. But with both eyes open, you gain the depth perception of a *two-point* perspective —the difference between the two perspectives being the result of a slight rotation around the vertical axis (the *y*-axis on the HP 48). Or, if you combine the rotation around the *y*-axis with the rotation around the *x*-axis, you can get a *three-point*, or *oblique*, perspective.

In terms of transformation matrices, the two-point and three-point perspectives are nothing more than the combination (multiplication) of one or two rotation matrices with the single-point perspective transformation matrix. The resulting combined transformation matrices (with θ the angle of rotation around the *y*-axis and ϕ the angle of rotation around the *x*-axis) are:

2-pt:
$$\begin{bmatrix} \cos\theta & 0 & 0 & \frac{-\sin\theta}{k} \\ 0 & 1 & 0 & 0 \\ \sin\theta & 0 & 0 & \frac{\cos\theta}{k} \\ 0 & 0 & 0 & 1 \end{bmatrix}$$

3-pt:
$$\begin{bmatrix} \cos\theta & \sin\theta\sin\phi & 0 & \frac{-\sin\theta\cos\phi}{k} \\ 0 & \cos\phi & 0 & \frac{\sin\phi}{k} \\ \sin\theta & -\cos\theta\sin\phi & 0 & \frac{\cos\theta\cos\phi}{k} \\ 0 & 0 & 0 & 1 \end{bmatrix}$$

Of course, a program makes perspective projections much easier: PERSP (see page 292) takes an object array from level 3, the desired translation vector from level 2, and the desired eyepoint from level 1. It returns the object array ready for TVIEW.

The translation vector allows you to move the entire object with respect to the origin. This lets you treat the eyepoint value as relative to the object (as well as to the origin), which makes it much easier to anticipate the perspective you obtain.

Whether you get a one-, two-, or three-point perspective depends on your choice of eyepoint. If only the z-coordinate is non-zero, the perspective is single-point; if the y-coordinate is also non-zero, the perspective is two-point; if all three co-ordinates are non-zero, the perspective is three-point. Be sure to use a negative number for the eyepoint's z-coordinate so as not to view the object from *inside* it.

Try the next few examples to get a feel for perspective projections.

Example: Project the CUBE using no translation ([0 0 0]) and an eyepoint of [2 0 -10]. This produces a two-point perspective.

1. Return to the stack, drop the previous result array, and put CUBE onto the stack as the object array: (CANCEL)(◀)(VAR) CUBE .

2. Enter the translation vector: (⟵)[ ] 0 (SPC) 0 (SPC) 0 (ENTER).

3. Enter the eyepoint vector: (⟵)[ ] 2 (SPC) 0 (SPC) 1 0 (+/−)(ENTER).

4. Find the perspective projection (with the normalization: (α)(α)(P)(E) (R)(S)(P)(ENTER) or (VAR) (then (NXT) or (⟵)(PREV) as needed) PERSP .

5. Display the results using TVIEW: TVIEW .

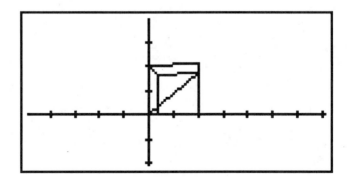

Example: Project the CUBE using the same eyepoint as the previous example, but this time translate the cube to [4 5 4].

1. Return to the stack, drop the previous result array, and put CUBE onto the stack as the object array: [CANCEL] [←] [VAR] **CUBE** .

2. Enter the translation vector: [←] [[ ]] [4] [SPC] [5] [SPC] [4] [ENTER].

3. Enter the eyepoint vector: [←] [[ ]] [2] [SPC] [0] [SPC] [1] [0] [+/-] [ENTER].

4. Find the perspective projection (with the normalization): [α] [α] [P] [E] [R] [S] [P] [ENTER] or [VAR] (then [NXT] or [←] [PREV] as needed) **PERSP**.

5. Display the results using TVIEW: **TVIEW**.

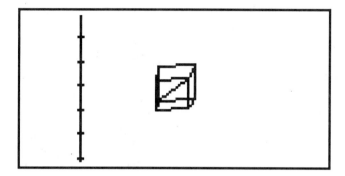

Note that the perspective in this case looks at the cube from below it and to the left, visually skewing it accordingly. This is the effect of the relationship of the eyepoint to the center of the object.

Example: Project CUBE using a [-1 -1 0] translation vector and a [6 6 -5] eyepoint. This produces a three-point perspective.

1. Return to the stack, drop the previous result array, and put CUBE onto the stack as the object array: CANCEL ← VAR **CUBE** .

2. The translation vector: ← [ ] 1 +/− SPC 1 +/− SPC 0 ENTER .

3. Enter the eyepoint vector: ← [ ] 6 SPC 6 SPC 5 +/− ENTER .

4. Find the perspective projection (with the normalization): α α P E R S P ENTER or VAR (then NXT or ← PREV as needed) **PERSP** .

5. Display the results using TVIEW: **TVIEW** .

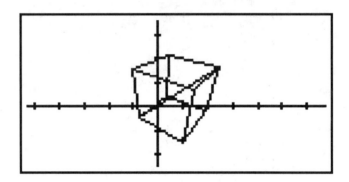

As a curiosity and illustration of how deceptive a view it can be, try one final example where the eyepoint is located *inside the object*.

Example: Project the CUBE using a [-1 -1 0] translation vector and [0 0 1] as the eyepoint. This gives an insider's perspective!

1. Return to the stack, drop the previous result array, and put CUBE onto the stack as the object array: CANCEL ← VAR CUBE .

2. The translation vector: ← [ ] 1 +/− SPC 1 +/− SPC 0 ENTER .

3. Enter the eyepoint vector: ← [ ] 0 SPC 0 SPC 1 ENTER .

4. Find the perspective projection (with the normalization): α α P E R S P ENTER or VAR (then NXT or ← PREV as needed) PERSP .

5. Display the results using TVIEW: TVIEW .

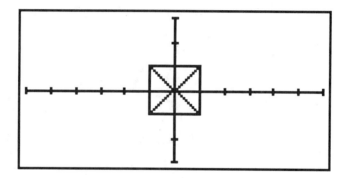

7. Conic Sections

Introduction to Conic Sections

A *conic section* is a widely used form of plane curve that is defined in any of three equivalent (and interchangeable) ways:

- It is formed by the set of all points in a plane whose distances from a fixed point (*focus*) divided by their distances from a fixed line (or *directrix*) is a constant ratio, ε (or *eccentricity*).

- It is the result of the general second-degree algebraic curve:

$$Ax^2 + Bxy + Cy^2 + Dx + Ey + F = 0$$

- It is formed from the intersection of a plane and a right circular double cone —a "cross-section of a cone" (hence the name). There are four general shapes of conic sections, depending on: the angle (α) made by the intersecting plane with respect to the bases of the cones; and the angle (β) made by the cone itself with respect to its base.

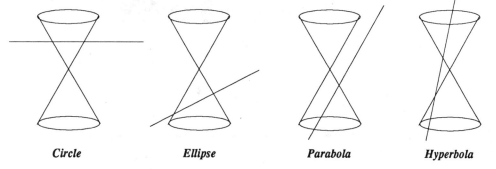

| *Circle* | *Ellipse* | *Parabola* | *Hyperbola* |

- – If the intersecting plane is parallel to the bases (that is, α = 0), the cross section is a *circle* (unless the intersection is at the cone's vertex only, in which case the cross section reduces to a single point).

- – If 0 < α < β, the cross section is an *ellipse* (unless the intersection is at the cone's vertex only, in which case it reduces to a single point).

- – If α = β, the cross section is a *parabola* (unless the plane contains the cone's vertex, in which case the cross section is a straight line).

- – If α > β, the cross section is a *hyperbola* (unless the plane contains the cone's vertex, in which case the cross section is a pair of intersecting lines).

Plotting Conics

This chapter illustrates how the three different descriptions fit together for each kind of conic section. You will see how to rotate and translate the conics, how to compute various analytical quantities, and how to convert between descriptions. But before looking at each of the four conic types, look at the HP 48's specialized plot type, `Conic`. This plot type will plot any implicit function of two real variables which is of no more than second order in either variable. *So, in fact, it will plot many implicit functions that don't produce conic sections.* Some examples:

$\sin x^2 - \cos y - 1 = 0$:

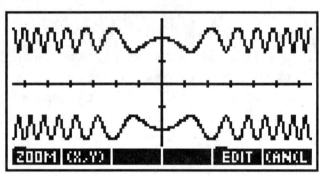

$x - y \log x^2 - xy = 0$:

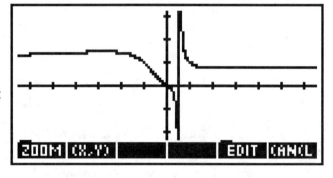

$4xy^2 - y^3 - x = 0$:

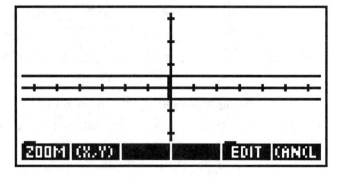

There are good uses for such generality,* but nevertheless you will use the **Conic** plot type primarily to plot conic sections—and hence the name.

Technically, most conic sections are *not functions* (because each input can yield more than exactly one output). So, essentially, the **Conic** plot type breaks the second-degree equation into two equivalent true functions and plots each with the **Function** plot type—showing you both plots simultaneously.

Example: Use **Conic** to plot this circle: $(x-1)^2 + y^2 = 4$

1. Open the **PLOT** application, highlight the **TYPE:** field and change the plot type to **Conic**.

2. Reset the plot parameters: (DEL)(▼)(ENTER).

3. In the **EQ:** field, enter the circle's equation: (▼)(←)(EQUATION)(←)(( ))(α)(←)(X)(−)(1)(▶)(y^x)(2)(▶)(+)(α)(←)(Y)(y^x)(2)(▶)(←)(=)(4)(ENTER).

4. Change the **INDEP:** variable to **X** (lower-case) and the **DEPND:** variable to **y** (lower-case). You will find the **DEPND:** setting in the **PLOT OPTIONS** screen (press **OPTS** from the main **PLOT** screen).

5. Leave all other plot options at their default settings and draw the plot, returning first to the main plot screen, if necessary: (**OK** , if needed) **ERASE** **DRAW**. Note how the plot is drawn in two pieces simultaneously, just as if you were plotting two functions simultaneously.

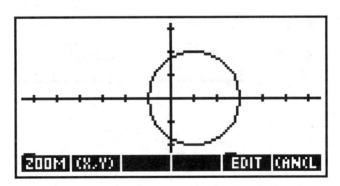

*Actually, **Conic** will plot *any* implicit function of two real variables, regardless of order, as long as it can compute a second-order Taylor's approximation of the function. So **Conic** plots of two-variable polynomials with an order higher than two in either variable will be approximations but are often adequate for plotting purposes.

Sometimes you will need to adjust the step size to see the entire conic clearly.

Example: Plot the following conic, using default settings: $x^2 + 3y^2 = 6$

1. Return to the **PLOT** screen: (CANCEL).
2. Highlight the **EQ:** field and enter the conic: (←)(EQUATION)(α)(←)(X)
   (y^x)(2)(▶)(+)(3)(×)(α)(←)(Y)(y^x)(2)(▶)(←)(=)(6)(ENTER).
3. Draw the plot: **ERASE DRAW**.

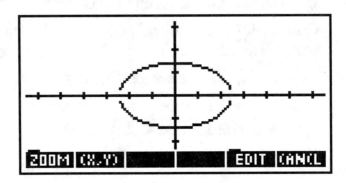

Parts of the ellipse are not fully drawn. Correct this by decreasing the step size....

Example: Repeat the above example with a step size of 0.02.

1. Return to the **PLOT OPTIONS** screen: (CANCEL) **OPTS**.
2. Highlight the **STEP:** field and enter .02: (▼)(▼)(·)(0)(2)(ENTER).
3. Redraw the plot: **OK ERASE DRAW**.

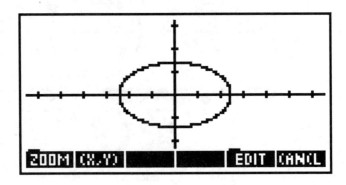

The other concern about plotting conic sections involves the display range: If you allow the scale of the two axes to vary from one another, the image may be distorted in a misleading way.

Example: Plot the circle $x^2 + y^2 = 4$, using a square viewing area.

1. Return to the **PLOT** screen and highlight the **EQ:** field.

2. Enter the equation for the circle: ⬅EQUATION α ⬅ X y^x 2 ▶ + α ⬅ Y y^x 2 ▶ ⬅ = 4 ENTER.

3. Make sure that the **INDEP:** variable is ✕ (lower-case) and that the **DEPND:** variable is ੫ (lower-case).

4. Make the display ranges for **H-VIEW** and **V-VIEW** identical. Set both to ‾3 3.

5. Plot the circle: **ERASE DRAW**.

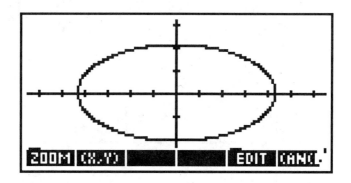

The plot looks more like an ellipse than a circle, because the display range doesn't match the shape of the display itself—the circle has been stretched to accommodate the square coordinates you requested. That is, the horizontal and vertical ranges are identical and yet there are roughly twice the number of pixels horizontally as vertically, so each pixel on the horizontal axis represents about 0.05 units, while each pixel on the vertical axis represents about 0.1 units.

Moral of the Story: With a display width roughly twice that of its height, when plotting conic sections—particularly circles and ellipses—you should set the horizontal display range to roughly twice that of the vertical range in order to get a plot that isn't visually distorted.

The program CONPLT (see page 278) streamlines the plotting of conic sections so that they appear well-centered and undistorted. CONPLT takes a list of the six coefficients of the conic section in general form from level 1 and plots the conic, returning nothing to the stack. Thus, given the general form of a conic,

$$Ax^2 + Bxy + Cy^2 + Dx + Ey + F = 0$$

CONPLT takes as its only input a list of the coefficients in the order shown above: { A B C D E F }. Note that you must enter a zero as the coefficient for any missing term—that is, the input list must have exactly six entries.

The following two examples illustrate the use of CONPLT.

Example: Use CONPLT to plot the conic $4x^2 + 3xy - 5y^2 - 2x + y - 25 = 0$

1. Enter the list of coefficients for the conic onto the stack: CANCEL ← { }
   4 SPC 3 SPC 5 +/− SPC 2 +/− SPC 1 SPC 2 5 +/− ENTER.

2. Plot the conic using CONPLT: α α C O N P L T ENTER or VAR
   (then NXT or ← PREV as needed) CONP.

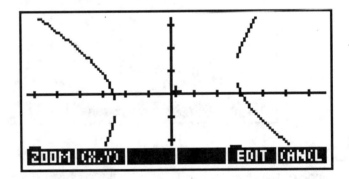

This conic is a hyperbola that is somewhat rotated with respect to the coordinate axes.

Example: Use CONPLT to plot the conic $(x-4)^2 + (y+2)^2 = 25$.

1. Because the conic isn't in general form, you must first expand the left-hand side, collect terms and move all terms to the left-hand side.
   Result: $x^2 + y^2 - 8x + 4y - 5 = 0$

2. Enter the list of coefficients for the conic onto the stack: [←][{ }][1] [SPC][0][SPC][1][SPC][8][+/−][SPC][4][SPC][5][+/−][ENTER].

3. Plot the conic using CONPLT: [α][α][C][O][N][P][L][T][ENTER] or [VAR] (then [NXT] or [←][PREV] as needed) **CONP**.

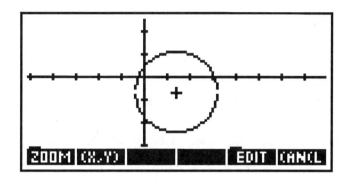

Circles

The general conic equation becomes that of a *circle* when $A = C$ and $B = 0$:

$$Ax^2 + Ay^2 + Dx + Ey + F = 0$$

Defined geometrically, a circle is the set of coplanar points equidistant from a given fixed point (the center). Viewed as such, the circle has two defining parameters: its center, (h,k); and its radius, r, related by $(x-h)^2 + (y-k)^2 = r^2$

The two programs CIR→G and G→CIR (see pages 277 and 282, respectively) convert between the center-radius form of the equation and the general form of the equation of a circle. The center-radius form is given by a complex number on level 2 (representing the coordinates of the center) and a real number on level 1 (representing the radius). The general form is given as a list of the six coefficients of the general conic equation, which, for a circle will be { A 0 A D E F }.

Example: The general equation of a circle centered at (2, -3), with radius 7, is?

1. Enter the center of the circle as a complex number: ⬅(()) 2 SPC 3 +/- ENTER.

2. Enter the radius: 7 ENTER.

3. Find the general equation via CIR→G: α α C I R ↱ ← G ENTER or VAR (then NXT or ⬅PREV as needed) CIR→G.

    Result: { 1 0 1 '-4' 6 '-36' }

    So the general equation is $x^2 + y^2 - 4x + 6y - 36 = 0$.

Example: Find the center and radius of the circle $4x^2 + 4y^2 - x + 5y - 3 = 0$

1. Enter the general equation as a list of coefficients: ⬅{} 4 SPC 0 SPC 4 SPC 1 +/- SPC 5 SPC 3 +/- ENTER.

2. Compute the center and radius using G→CIR: α α G ↱ ← C I R ENTER or VAR (then NXT or ⬅PREV as needed) G→CIR.

    Result: 2:   (.125, -.625)
            1:      1.07529065838

    So the center of the circle is $\left(\dfrac{1}{8}, -\dfrac{5}{8}\right)$, and the radius is ≈ 1.075.

There are several important relationships involving circles that you may remember from a geometry class. Here's a brief review:

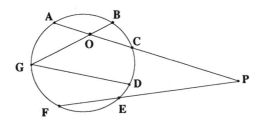

- Three noncollinear points in plane determine a unique circle.
- The measure of an inscribed angle is one-half the measure of its intercepted arc. For example, in the above diagram: $m\angle BGD = \dfrac{1}{2}\text{arc}(BD)$.
- If two chords intersect in the interior of a circle, the measure of the angle formed is the average of the measures of the arcs intercepted by the angle and its opposite or vertical angle. For example, in the above diagram:

$$m\angle BOC = m\angle AOG = \dfrac{1}{2}(\text{arc}(BC) + \text{arc}(AG))$$

- If two secants intersect in the exterior of a circle, the measure of the angle formed is one half the difference of the measures of the intercepted arcs.

  For example, in the above diagram: $m\angle APF = \dfrac{1}{2}(\text{arc}(AF) - \text{arc}(CE))$

- If two chords intersect inside a circle, then the products of the lengths of the segments of each chord are equal. For example, in the above diagram: $(BO)(OG) = (AO)(OC)$

- If two secant segments are drawn to a circle from the same exterior point, the product of the lengths of one secant segment and its external segment equals the product of the lengths of the other secant segment and its external segment. For example, in the above diagram: $(AP)(CP) = (FP)(EP)$

Using the first of these relationships suggests that you should be able to compute the equation of a circle, given three noncollinear points. You can.

- *Method 1:* Find the center of the circle by finding the intersection of the perpendicular bisectors of the segments connecting the points; compute the radius by finding the distance from the center to any one of the points.

- *Method 2*: Replace the x and y variables in the general form of the circle equation with each of the three points (and let $A = 1$), to get a linear system.

The following two examples illustrate each of these methods:

Example: Find the equation of the circle through the three points $R(1,0)$, $S(0,1)$ and $T(2,2)$ using the perpendicular bisector method.

1. Enter points R and S onto the stack in vector form: ⬅[] 1 SPC 0 ENTER ⬅[] 0 SPC 1 ENTER.

2. Compute the perpendicular bisector of the segment RS using the program P2→PB (introduced on page 184 in Chapter 6): VAR (then NXT or ⬅PREV as needed) **P2→P**.

3. Convert the perpendicular bisector to array form (via the programs I→GEN and G→A from Chapter 6): Press (NXT or ⬅PREV as needed) **I→GE** SWAP ⬅ **G→A**.

4. Using points S and T, repeat steps 1 - 3: ⬅[] 0 SPC 1 ENTER ⬅ [] 2 SPC 2 ENTER **P2→P** **I→GE** SWAP ⬅ **G→A**.

5. Find the intersection point of the two perpendicular bisectors, using LIN2? (see page 200): Press (NXT or ⬅PREV as needed) **LIN2?** ⬅. Result: [1.16667 1.16667] (to 5 decimal places).

6. Copy the result, enter one point, and find the radius: ENTER ⬅[] 1 SPC 0 ENTER ⊖ MTH **VECTR** **ABS**. Result: 1.17851

7. Swap the center point into level 1 and convert it to a complex number: SWAP PRG **LIST** **OBJ→** ⬅ MTH NXT **CMPL** **R→C**.

8. Swap and find the equation: SWAP VAR (NXT or ⬅PREV)) **CIR→G**. Result: { 1 0 1 '-(7/3)' '-(7/3)' '4/3' }

Thus the equation of the circle is $x^2 + y^2 - \dfrac{7}{3}x - \dfrac{7}{3}y + \dfrac{4}{3} = 0$

Example: Find the equation of the circle through the three points, $A(2,3), B(3,-1)$ and $C(-2,1)$ using the linear systems method.

1. Create the system of three equations in three unknowns by substituting each of the given points into the general equation for a circle:

$$2D + 3E + F = -13$$
$$3D - E + F = -10$$
$$-2D + E + F = -5$$

2. Open the $\mathsf{Solve\ lin\ sys}_\mathsf{...}$ application; enter the matrix of coefficients: [→][SOLVE][▲][▲][ENTER][→][MATRIX][2][ENTER][3][ENTER] [1][ENTER][▼][3][ENTER][1][+/−][ENTER][1][ENTER][2][+/−][ENTER][1] [ENTER][1][ENTER][ENTER].

3. Highlight the **B:** field and enter the vector of constants: [▼][←][[ ]] [1][3][+/−][SPC][1][0][+/−][SPC][5][+/−][ENTER].

4. Solve the linear system: **SOLVE**.

   Result: $\mathtt{[\ -1.44444\ -1.11111\ -6.77778\ ]}$

5. *Optional.* Although you have already computed the three missing coefficients, you can put them into proper general form by converting the resulting vector to rational values using $\mathsf{R{\to}Q}$ and prepending the first three coefficients, { 1 0 1 }: [CANCEL][EVAL][VAR](then [NXT] or [←][PREV] as needed) **R→Q** [←][{ }][1][SPC][0][SPC][1][ENTER] [SWAP][+].

   Result: $\mathtt{\{\ 1\ 0\ 1\ '{-}(13/9)'\ '{-}(10/9)'}$
   $\mathtt{'{-}(61/9)'\ \}}$

6. *Optional.* Make the coefficients integral by multiplying through by 9 and collecting: [9][×][←][SYMBOLIC] **COLCT**.

   Result: $\mathtt{\{\ 9\ 0\ 9\ -13\ -9.99999999999\ -61\ \}}$

As in this case, you may see a round-off error when creating integral coefficients. So the circle is $9x^2 + 9y^2 - 13x - 10y - 61 = 0$.

Points and Circles

Given a circle with center at (h, k) and a radius r, and a point (x,y), there is an easy way to determine whether or not the point lies on the interior of the circle, exterior of the circle, or on the circle itself:

- If $(x - h)^2 + (y - k)^2 < r^2$, then the point lies on the interior of the circle.
- If $(x - h)^2 + (y - k)^2 = r^2$, then the point lies on the circle itself.
- If $(x - h)^2 + (y - k)^2 > r^2$, then the point lies on the exterior of the circle.

Example: Does the point (-1, 2) lie in the exterior of, in the interior of, or on the circle $3x^2 + 3y^2 - 4x + 6y - 10 = 0$?

1. Enter the circle in general form (as a list of coefficients): (←)[{ }] 3 SPC 0 SPC 3 SPC 4 +/- SPC 6 SPC 1 0 +/- ENTER.

2. Convert it to center-radius form: VAR G→CIR.

3. Square the radius: (←) X².

4. Swap the center into level 1, enter the point as a complex number, and subtract from the center. Note that this is essentially the same as finding the vector between the center and the point: SWAP (←) ( ) 1 +/- SPC 2 ENTER −.

5. Find the square of the absolute value of the previous result: MTH NXT CMPL ABS (←) X². Result: 11.777778.

6. Compare this result with the previous one (on level 2). Level 1 is larger, indicating that the point lies outside of the circle.

Ellipses

An *ellipse* is the set of points in a plane whose distances from two fixed points in the plane have a constant sum. An ellipse is a more general version of a circle in that it has two axes of different lengths—the *major* and *minor* axes, the major axis being the longer of the two—rather than a single radius.

The ellipse has four parameters:

- The *center* (h,k) located at the midpoint joining the two fixed points, or *foci*, defining the ellipse.

- The *semimajor* (a)—half of the length of the major axis.

- Either the *semiminor* (b)—half of the length of the minor axis—or the *eccentricity* (e), a ratio of the distance from the center to either foci compared with the semimajor. Either of these parameters will do because they are related to each other by the following: $b^2 = a^2(1 - e^2)$

- The *angle of orientation* (θ) between the major axis and the *x*-axis.

The standard equation for an ellipse assumes that the angle of orientation is zero:

$$\frac{(x-h)^2}{a^2} + \frac{(y-k)^2}{b^2} = 1$$

However, the general equation for an ellipse makes no assumption about the angle of orientation and is used whenever there is some rotation of the ellipse:

$$Ax^2 + Bxy + Cy^2 + Dx + Ey + F = 0$$

where neither A nor C is zero, $A \neq C$, and $AC > 0$. The angle of orientation is:

$$\theta = \frac{1}{2}\tan^{-1}\left(\frac{B}{A-C}\right)$$

Note that whenever the B-coefficient is zero, the angle of orientation is zero, and so the standard equation can be used as well.

The programs G→ELP and ELP→G (see pages 283 and 281, respectively) convert between the general equation of an ellipse and the set of four parameters: center, semimajor, semiminor, and angle of orientation. Here are some examples....

Example: Find and plot the general equation of an ellipse centered at (-2,3), with semiaxes of 5 and 3 and an angle of orientation is 30°.

1. Make sure that you're in degree mode (⬅️[RAD], if necessary), and enter the center of the ellipse: ⬅️[( )][2][+/−][SPC][3][ENTER].

2. Enter the list of parameters: ⬅️[{ }][5][SPC][3][SPC][3][0][ENTER].

3. Find the coefficients of the equation: [α][α][E][L][P][➡️][⬅️][G][ENTER] or [VAR] (then [NXT] or ⬅️[PREV] as needed) **ELP+**. Result:

   { .6190 .6598 1 -2.0868 -7.0430 1.7143 }

4. Plot the ellipse: ([NXT] or ⬅️[PREV] as needed) **COMP**.

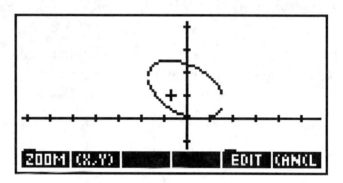

Example: Find the center, semimajor, and eccentricity of the following ellipse:

$$25x^2 + 9y^2 - 100x + 54y - 44 = 0$$

1. Enter the ellipse as a list of coefficients: [CANCEL]⬅️[{ }][2][5][SPC][0] [SPC][9][SPC][1][0][0][+/−][SPC][5][4][SPC][4][4][+/−][ENTER].

2. Find the ellipse parameters: [α][α][G][➡️][⬅️][E][L][P][ENTER] or [VAR] ([NXT] or ⬅️[PREV] as needed) **G+EL**. Result: 2:   (2,-3)
                                                                    1: { 3 5 0 }

   The center of the ellipse is (2,-3) and the semimajor is 5—the larger of the first two elements in the level 1 list.

3. Compute the eccentricity. It is the square root of 1 minus the square of the ratio of semiminor to semimajor: [PRG] **LIST** **OBJ+** [◀][◀][÷] ⬅️[x²][1][−][+/−][√x]. Result: .8

In the introduction to this chapter (page 243), the first definition of conic sections given was that of a planar curve formed by the set of all points such that, for each point, its distance from a fixed point (the *focus*) divided by its distance from a fixed line (the *directrix*) is a constant ε (the *eccentricity*).

The ellipse actually has two foci and two directrixes, as the diagram shows below:

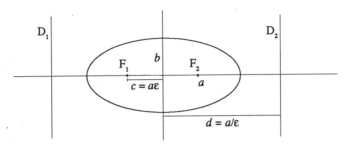

The foci F_1 and F_2 are each a distance equal to the product of the semimajor, a and the eccentricity, ε. The directrixes, D_1 and D_2, are parallel to the minor axis and are a distance equal to the quotient of the semimajor and the eccentricity.

Example: Find the coordinates of the vertices and foci of the ellipse in the previous example: $25x^2 + 9y^2 - 100x + 54y - 44 = 0$

1. Find the basic parameters for the ellipse. From the results in the previous example, you know that the center is located at (2,-3), the semimajor is 5, the eccentricity is 0.8, and the major axis runs vertically (i.e. parallel to the *y*-axis).

2. The vertices are located along the major axis at a distance of 5 on either side of the center. Thus, vary the y-coordinate by ±5: The vertices are (2, 2) and (2,-8).

3. The foci are also on the major axis at a distance equal to the product of the semimajor and the eccentricity. Thus the y-coordinate must be adjusted by (5)(0.8) or ±4: The foci are (2,1) and (2,-7).

Example: Find the equation of an ellipse with an eccentricity of 2/3 and the line $x = 9$ is one directrix with its corresponding focus at (4,0).

1. Analyze the given information. The directrix is a vertical line, so the major axis is horizontal. The focus is located on the x-axis, so the major axis is the x-axis. If the vertex is a distance a from the center $(h,0)$, then the distance from the directrix to the center is a/e or $1.5a$ and the distance from the focus to the center is ae or $.67a$.

2. This leads to two equations in two variables: $1.5a = 9 - h$
$$(2/3)a = 4 - h$$

   Set both equations equal to h and solve for a, then backsolve for h. Result: $a = 6$; $h = 0$

3. Enter the coordinates of the center: ⬅ ( ) 0 SPC 0 ENTER.

4. Enter a, make a copy, and enter the eccentricity: 6 ENTER ENTER 2 ENTER 3 ÷.

5. Compute the semiminor, b: ⬅ x² 1 ─ +/- SWAP ⬅ x² × √x. Result: **4.472135955.**

6. Enter 0 as the angle of orientation and assemble the parameters into a list: 0 ENTER 3 PRG **LIST →LIST** .

7. Find the coefficients of the ellipse: VAR (NXT) or ⬅ PREV as needed) **ELP→G**. Result: **{ .555555556 0 1 0 0 -20 }.**

   Multiplying through by 9 to make the coefficients all integers gives the ellipse $5x^2 + 9y^2 - 180 = 0$

Example: Find the eccentricity and directrixes of the ellipse $\dfrac{x^2}{7} + \dfrac{y^2}{16} = 1$

1. Analyze the ellipse. The center is the origin and the major axis is the y-axis (because b^2, 16, is greater than a^2, 7).

2. Enter the eccentricity: 1 ENTER 7 ENTER 1 6 ENTER ÷ ─ √x. Result: **.75**

3. The directrixes are $y = \pm b/e$. So, compute b/e: 1 6 √x SWAP ÷.

   Result: **5.333.** Thus the directrixes are $y = \pm\dfrac{16}{3}$

Parabolas

A *parabola* is the set of points in a plane that are equidistant from a given fixed point (*focus*) and fixed line (*directrix*) in the plane. The eccentricity of a parabola is always 1. The *vertex* is the point of the parabola closest to the directrix.

Parabolas are controlled by three parameters:

- The location of the vertex (h,k).
- The signed distance between the focus and the vertex (p).
- The *angle of orientation* (θ)—the angle between the axis of symmetry and the appropriate reference axis (either the y- or x-axis).

The standard form of the equation of a parabola with an axis of symmetry parallel to the y-axis and with its vertex at (h,k) is $(x-h)^2 = 4p(y-k)$. If the parabola has an axis of symmetry parallel to the x-axis, the equation is $(y-k)^2 = 4p(x-h)$. The absence of the second-degree term in either x or y (but not both) is a characteristic of the parabola. Indeed, any second-degree polynomial of one variable defines a parabola.

A conic given in general form, $Ax^2 + Bxy + Cy^2 + Dx + Ey + F = 0$, is a parabola if $B^2 - 4AC = 0$. The general form is used whenever the angle of orientation is nonzero.

The programs G→PBL and PBL→G (see pages 283 and 290, respectively) convert between the general equation and the set of three parameters. G→PBL takes the list of coefficients representing a general conic from level 1 and returns the coordinates of the vertex to level 2 and a list containing the p parameter and the angle of orientation to level 1. The PBL→G takes a complex number representing the vertex from level 2 and a two-element list containing the p parameter and the angle of orientation (in degrees) from level 1 and returns the list of general conic coefficients to level 1.

Example: Find the focus of the parabola $2x^2 - 3x + 5y + 4 = 0$

1. Enter the parabola in general form: ⬅️{ } ② SPC ⓪ SPC ⓪ SPC ③ +/− SPC ⑤ SPC ④ ENTER.

2. Find the parameters: α α G →P B L ENTER or VAR (NXT or ⬅️ PREV as needed) **G→PBL**. Result: 2:   (.75, -.575)
   1:    { -.625 0 }

3. The parabola has an axis parallel to the y-axis and, because the p parameter is negative, it opens downward. Thus the y-coordinate of the focus differs by p (it will be more negative) from that of the vertex. Thus the focus is (.75,-1.2).

Example: Find the general equation of the parabola with the vertex at (-2,-2) and the line $y = -3$ as its directrix.

1. Enter the vertex: ⬅️( ) ② +/− SPC ② +/− ENTER.

2. The directrix is horizontal and below the vertex: the parabola opens upwards; $p = -2--3 = 1$. The parameter list: ⬅️{ } ① SPC ⓪ ENTER.

3. For the general equation: α α P B L →G ENTER or VAR NXT or ⬅️ PREV as needed) **PBL→**. Result: { 0 0 1 -4 4 -4 }

Note that, by default, the PBL→G program assumes that the parabolic axis is parallel to the x-axis (i.e. that it's second-degree in y). To convert it to the parabola with the axis parallel to the y-axis, swap the A- and C-coefficients with each other and the D- and E-coefficients with each other, making it { 1 0 0 4 -4 -4 }. Thus the equation for the parabola is $x^2 + 4x - 4y - 4 = 0$

Hyperbolas

A *hyperbola* is the set of points in a plane whose distances from two fixed points in the plane have a constant difference. A hyperbola has two foci, located a distance c on either side of the center, along the main axis of the hyperbola. Associated with each focus is a vertex, located at a distance a on either side of the center.

The hyperbola has four parameters:

- The *center* (h,k) located at the midpoint joining the two fixed points, or *foci*, defining the hyperbola.

- The *distance between center and each vertex* (a).

- Any one of the following:

    - The *distance between center and each focus* (c).

    - The *parameter* (b), computed as $b = \sqrt{c^2 - a^2}$.

    - The *eccentricity* (e), equal to the ratio $\dfrac{c}{a}$ and to $\sqrt{1 - \dfrac{b^2}{a^2}}$.

- The *angle of orientation* (θ) between the main axis and the x-axis.

The standard hyperbola equation assumes that the angle of orientation is zero:

$$\frac{(x-h)^2}{a^2} - \frac{(y-k)^2}{b^2} = 1$$

However, the general equation for a hyperbola makes no assumption about the angle of orientation and is used whenever there is some rotation of the hyperbola:

$$Ax^2 + Bxy + Cy^2 + Dx + Ey + F = 0$$

where $B^2 - 4AC > 0$. The angle of orientation is $\theta = \dfrac{1}{2}\tan^{-1}\left(\dfrac{B}{A-C}\right)$.

Note that whenever the B-coefficient is zero, the angle of orientation is zero and then the standard equation can be used as well.

The programs G→HYP and HYP→G (see pages 283 and 284, respectively) convert between the general equation of a hyperbola and the set of 4 parameters: center, the a parameter, b parameter, and orientation angle. Here are some examples.

HYP→G takes the complex number representing the center from level 2, and a list containing, in order, the *a* parameter, the *b* parameter and angle (in degrees) of orientation, and converts it to a list of the six coefficients of the general equation. G→HYP does the reverse conversion.

Example: Find and plot the general equation of a hyperbola centered at (-2,3), with an *a* of 5, a *b* of 3, and angle of orientation of 30°.

1. Enter the center: [←][( )][2][+/-][SPC][3][ENTER].
2. Enter the list of parameters: [←][{ }][5][SPC][3][SPC][3][0][ENTER].
3. The general equation: [α][α][G][→][→][H][Y][P][ENTER] or [VAR] ([NXT] or [←][PREV] as needed) HYP→G. Result (to 5 places): `{ .03030 -1.78454 -1 6.43496 6.78205 -25.09091 }`

   The hyperbola: $0.03x^2 - 1.78xy - y^2 + 6.43x + 6.78y - 25.09 = 0$
4. Plot the conic: [VAR] (then [NXT] or [←][PREV] as needed) CONP.

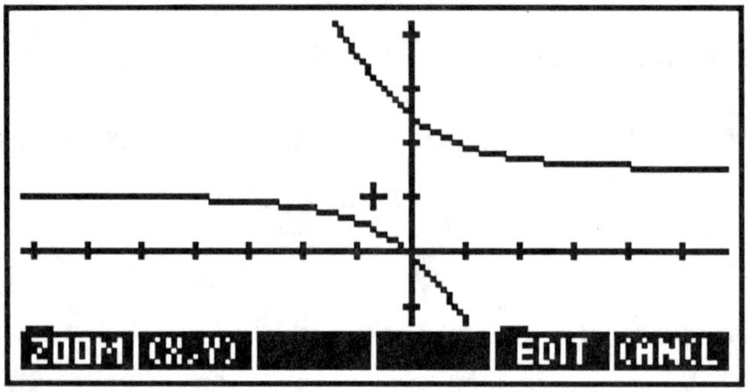

Example: Find the center and eccentricity of the following hyperbola:

$$25x^2 - 9y^2 - 100x + 54y - 44 = 0$$

1. Enter the hyperbola as a list of coefficients: [←][{ }][2][5][SPC][0][SPC][9][+/-][SPC][1][0][0][+/-][SPC][5][4][SPC][4][4][+/-][ENTER].
2. Compute the parameters: [α][α][G][→][H][Y][P][ENTER] or [VAR] (then [NXT] or [←][PREV] as needed) G→HYP. Result: `2: (2,3)` `1: { 3 5 0 }`
3. The eccentricity: [3][←][x²][5][←][x²][+][√x][3][/]. Result: `1.944`

In the introduction to this chapter (page 243), the first definition of conic sections given was that of a planar curve formed by the set of all points such that, for each point, its distance from a fixed point (the *focus*) divided by its distance from a fixed line (the *directrix*) is a constant ε (the *eccentricity*).

Like the ellipse, the hyperbola has two foci and two directrixes. Unlike the ellipse, the hyperbola has two discontinuous branches constrained by two *asymptotes*. For hyperbolas in standard orientation, the equations of the directrixes are

$x = h \pm \dfrac{a}{e}$, and the equations of the asymptotes are $bx \pm ay - (bh \pm ak) = 0$.

Example: Find the asymptotes and foci of the hyperbola

$$25x^2 - 9y^2 - 100x + 54y - 44 = 0$$

1. Enter the hyperbola as a list of coefficients and compute its parameters: ⟵{ } 2 5 SPC 0 SPC 9 +/− SPC 1 0 0 +/− SPC 5 4 SPC 4 4 +/− ENTER VAR (NXT or ⟵PREV as needed) G→HYP.

   Result:　2:　　　　　(2, 3)
   　　　　　1:　{ 3 5 0 }

2. Compute the eccentricity: 3 ⟵ X² 5 ⟵ X² + √x 3 /

   Result: 1.944

3. Find the directrixes: 3 ÷ 1/x ENTER 2 ENTER SWAP − SWAP 2 +.

   Result:　2:　　.457
   　　　　　1:　3.543

   Thus the directrixes are $x = .457$ and $x = 3.543$.

4. Find the asymptotes. Since $a = 3, b = 5, h = 2$, and $k = -3$, the asymptotes are $5x + 3y - 19 = 0$ and $5x - 3y - 1 = 0$.

Example: Find the equation of a hyperbola with an eccentricity of 1.3, the line $x = 9$ as one directrix, and the corresponding focus at (4,0).

1. Analyze the given information. Because the directrix is a vertical line, you know that the major axis is horizontal. Because the focus is located on the x-axis, you know that the major axis is the x-axis. If the vertex is a distance a from the center $(h,0)$, then the distance from the directrix to the center is a/e or $.769a$ and the distance from the focus to the center is ae or $1.3a$. The directrix $(x = 9)$ is between the center $(h,0)$ and focus (4,0) in a hyperbola, so $4 < 9 < h$.

2. These facts lead to two equations in two variables:

$$.769a = h - 9 \ \text{ or } \ h - .769a = 9$$
$$1.3a = h - 4 \ \text{ or } \ h - 1.3a = 4$$

Solve the set of equations simultaneously: ⟵ [ ] 9 SPC 4 ENTER ⟶ MATRIX 1 ENTER 1 . 3 +/− 1/x ENTER ▼ 1 ENTER 1 . 3 +/− ENTER ENTER ÷.
   Result: **[16.24638 9.42029]**

Thus $h \approx 16.25$ and $a \approx 9.42$.

3. Enter the coordinates of the center $(h,0)$: MTH **VECTR V→** SWAP 0 MTH NXT **CMPL R→C**.

4. Swap a into level one, make two copies and enter the eccentricity: SWAP ENTER ENTER 1 . 3 ENTER.

5. Compute the parameter b: ⟵ x² SWAP ⟵ x² × SWAP ⟵ x² − √x. Result: **7.82508**

6. Enter 0 as the angle of rotation, and assemble the parameters into a list: 0 ENTER 3 PRG **LIST →LIST**.

7. Compute the coefficients of the hyperbola: VAR (then NXT or ⟵ PREV as needed) **HYP→G**.
   Result: **{ .69 0 −1 −22.42 0 120.89 }**

Then multiplying through by 100 to make the coefficients all integers gives the hyperbola $69x^2 - 100y^2 - 2242x + 12089 = 0$.

Lines and Conics

Points of Intersection

A line and a conic that share the same plane have one of three possible relationships:

- The line intersects the conic in two points.
- The line is tangent to the conic—intersecting it in one point.
- The line doesn't intersect the conic at all.

The program LCON? (see page 285) determines the point(s) of intersection, if any, of a conic and a line. It takes the conic as a list of its general-form coefficients from level 2 and the line in slope-intercept form from level 1 and returns a list to level 1. The result list will have two, one, or zero points (expressed as complex numbers), depending upon the relationship of line and conic.

The following examples illustrate the use of LCON? with a variety of conics:

Example: Find the points, if any, where the line $y = 4x - 2$ intersects the circle $x^2 + y^2 - 25 = 0$.

1. Enter the circle in general form (as a list of coefficients): ⏴[{}] [1] [SPC] [0] [SPC] [1] [SPC] [0] [SPC] [0] [SPC] [2] [5] [+/−] [ENTER].

2. Enter the line: ['] [α] ⏴[Y] ⏴[=] [4] [X] [α] ⏴[X] [−] [2] [ENTER].

3. Find the points of intersection, if they exist, by using LCON?: [α] [α] [L] [C] [O] [N] [?] [ENTER] or [VAR] ([NXT] or ⏴[PREV] as needed) **LCON**.

   Result: { (-.736369678158, -4.94547871263)
               (1.67754614875, 4.710184595) }

   The line is a *secant*—intersecting the circle at the two points whose coordinates are listed on level 1.

Example: Find the points of intersection, if any, of the line $y = x - 4$ and the ellipse $16x^2 - 4xy + 9y^2 - 64x + 54y - 26 = 0$.

1. Enter the conic as a list of coefficients: [←][{ }][1][6][SPC][4][+/−][SPC] [9][SPC][6][4][+/−][SPC][5][4][SPC][2][6][+/−][ENTER].

2. Enter the line: [′][α][←][Y][←][=][α][←][X][−][4][ENTER].

3. Execute LCON?: [VAR] (then [NXT] or [←][PREV] as needed) **LCON**.

   Result: { (-1.0999109, -5.099910)
   (4.242768, .242768) }

Example: Find the points of intersection, if any, of the line $y = x - 4$ and the parabola $9y^2 - 64x + 54y - 26 = 0$.

1. Enter the conic as a list of coefficients: [←][{ }][0][SPC][0][SPC][9][SPC] [6][4][+/−][SPC][5][4][SPC][2][6][+/−][ENTER].

2. Enter the line: [′][α][←][Y][←][=][α][←][X][−][4][ENTER].

3. Execute LCON?: [VAR] (then [NXT] or [←][PREV] as needed)**LCON**.

   Result: { (-1.06956, -5.06956)
   (10.18068, 6.18068) }

Example: Find the points of intersection, if any, of the line $y = x - 4$ and the hyperbola $16x^2 - 4xy - 9y^2 - 64x + 54y - 26 = 0$.

1. Enter the conic as a list of coefficients: [←][{ }][1][6][SPC][4][+/−][SPC] [9][+/−][SPC][6][4][+/−][SPC][5][4][SPC][2][6][+/−][ENTER].

2. Enter the line: [′][α][←][Y][←][=][α][←][X][−][4][ENTER].

3. Execute LCON?: [VAR] (then [NXT] or [←][PREV] as needed) **LCON**.

   Result: { (-30.253019, -34.253019)
   (4.253019, .253019) }

Tangents and Normals

Often it is useful to compute the equation of the line that is *tangent* to a given conic at a given point. Or perhaps it is the equation of the *normal*—the line perpendicular—at the given point that you require. The relationship between tangent and normal is illustrated here with a circle:

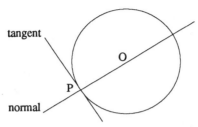

The program, TNCON (see page 314) computes the equations of the normal and tangent lines at a given point on a given conic.* TNCON accepts any conic in coordinate list form from level 2 and a point on the conic (as a complex number) from level 1 and returns labelled equations for the tangent and normal. Note that TNCON does not check to be sure that the given point actually lies on the conic and will give unreliable results if it doesn't.

The following examples involve tangents and normals of conics. Many, but not all, illustrate the use of TNCON.

Example: On the circle $4x^2 + 4y^2 - 3x - 6y - 17 = 0$, find equation of the tangent line through the point (-1,-1).

1. Enter the circle as a list of coefficients: ← { } 4 SPC 0 SPC 4 SPC 3 +/- SPC 6 +/- SPC 1 7 +/- ENTER .

2. Enter the point on the circle: ← ( ) 1 +/- SPC 1 +/- ENTER .

3. Execute TNCON: VAR (then NXT or ← PREV as needed) TNCON .

   <u>Result:</u> 2: Normal: 'y=-.90909+0.09091*x'
           1: Tangent: 'y=-12-11*x'

   Applying →Q to the normal equation reveals it as $y = -\dfrac{10}{11} + \dfrac{1}{11}x$.

---

*Note that TNCON uses differential calculus, a topic beyond the scope of this book. You may wish to wait to use TNCON until you've studied differential calculus a bit.

Conversely, you may also find the equation of a circle if you know the location of its center and a tangent. The only missing piece of information is the radius—which is nothing more than the distance from the center point to the tangent line.

Example: Find equation of circle centered at (-1, 1) that is tangent to the line, $x + 2y - 4 = 0$.

1. Enter the center point, make a copy, and convert it to a vector: ⬅️(·)(·) 1 +/− SPC 1 ENTER ENTER MTH NXT `CMPL` `C→R` MTH `VECTR` `→V2`. (Caution: If flag -19 is set, →V2 gives a complex number.)

2. Enter the line and convert to array form: ' ⬅️ α X + 2 X α ⬅️ Y − 4 ⬅️ = 0 ENTER VAR (then NXT or ⬅️ PREV as needed) `G→A`.

3. Find the distance to the line, using `DtoL` (see page 189 in Chapter 6): (NXT or ⬅️ PREV as needed) `DTOL`. Result: 1.341640786

4. Convert to a general equation for the circle—and clear any fractional coefficients—via multiplication: (use NXT or ⬅️ PREV as needed) `CIR→G` 5 X 1 ENTER ⬅️ « » ⬅️ →NUM ENTER PRG `LIST` `PROC` `DOLIS`. Result: { 5 0 5 10 −10 1 }.

Example: Find equation of the normal to the circle with a center at (2,-1) at the point (-1,3). Then find the equation of the circle itself.

1. Note that, for a circle, the normal for any point is a line containing both the point and the center of the circle. This fact allows you to compute the slope of the normal, which is the slope of the line containing the points given: ⬅️(·)(·) 1 +/− SPC 3 ENTER ⬅️(·)(·) 2 SPC 1 +/− ENTER − ENTER MTH NXT `CMPL` `C→R` SWAP ÷.
   Result: −1.33333333333

2. Compute the intercept of the normal: 1 +/− ENTER SWAP 2 X −.
   Result: 1.66666666666. Thus the normal is $y = -\dfrac{4}{3}x + \dfrac{5}{3}$.

3. Compute the radius of the circle by finding the absolute value of the difference between the position vectors of the center of the circle and the given point. Swap the copy of the difference vector into level 1 find its length: SWAP MTH `VECTR` `ABS`. Result: 5

4. Via the circle center, find its equation: $\boxed{\leftarrow}$ $\boxed{()}$ $\boxed{2}$ $\boxed{\text{SPC}}$ $\boxed{1}$ $\boxed{+/-}$ $\boxed{\text{ENTER}}$ $\boxed{\text{SWAP}}$ $\boxed{\text{VAR}}$ **CIR→C**. Result: $\{\ 1\ 0\ 1\ '-4'\ 2\ '-20'\ \}$
Thus, the equation of the circle is $x^2 + y^2 - 4x + 2y - 20 = 0$.

Example: Compute the equations of the lines normal and tangent to the ellipse
$16x^2 - 4xy + 9y^2 - 64x + 54y - 26 = 0$, at the point where $x = 1$ and
y is positive.

1. Enter the symbolic ellipse: $\boxed{'}$ $\boxed{1}$ $\boxed{6}$ $\boxed{\times}$ $\boxed{\alpha}$ $\boxed{\leftarrow}$ $\boxed{X}$ $\boxed{y^x}$ $\boxed{2}$ $\boxed{-}$ $\boxed{4}$ $\boxed{\times}$ $\boxed{\alpha}$ $\boxed{\leftarrow}$
$\boxed{X}$ $\boxed{\times}$ $\boxed{\alpha}$ $\boxed{\leftarrow}$ $\boxed{Y}$ $\boxed{+}$ $\boxed{9}$ $\boxed{\times}$ $\boxed{\alpha}$ $\boxed{\leftarrow}$ $\boxed{Y}$ $\boxed{y^x}$ $\boxed{2}$ $\boxed{-}$ $\boxed{6}$ $\boxed{4}$ $\boxed{\times}$ $\boxed{\alpha}$ $\boxed{\leftarrow}$ $\boxed{X}$ $\boxed{+}$ $\boxed{5}$ $\boxed{4}$ $\boxed{\times}$
$\boxed{\alpha}$ $\boxed{\leftarrow}$ $\boxed{Y}$ $\boxed{-}$ $\boxed{2}$ $\boxed{6}$ $\boxed{\text{ENTER}}$.

2. Store 1 into $'x'$ and solve for $'y'$: $\boxed{1}$ $\boxed{'}$ $\boxed{\alpha}$ $\boxed{\leftarrow}$ $\boxed{X}$ $\boxed{\text{STO}}$ $\boxed{'}$ $\boxed{\alpha}$ $\boxed{\leftarrow}$ $\boxed{Y}$
$\boxed{\text{ENTER}}$ $\boxed{1}$ $\boxed{0}$ $\boxed{\leftarrow}$ $\boxed{\text{SOLVE}}$ **ROOT** **ROOT**. Result: 1.21449871627

3. Convert the previous result into a complex number representing the
point of tangency: $\boxed{1}$ $\boxed{\text{ENTER}}$ $\boxed{\text{SWAP}}$ $\boxed{\text{MTH}}$ $\boxed{\text{NXT}}$ **CMPL** **R→C**.

4. Enter the ellipse as a list of coefficients: $\boxed{\leftarrow}$ $\boxed{\{\}}$ $\boxed{1}$ $\boxed{6}$ $\boxed{\text{SPC}}$ $\boxed{4}$ $\boxed{+/-}$
$\boxed{\text{SPC}}$ $\boxed{9}$ $\boxed{\text{SPC}}$ $\boxed{6}$ $\boxed{4}$ $\boxed{+/-}$ $\boxed{\text{SPC}}$ $\boxed{5}$ $\boxed{4}$ $\boxed{\text{SPC}}$ $\boxed{2}$ $\boxed{6}$ $\boxed{+/-}$ $\boxed{\text{ENTER}}$.

5. Compute the tangent and normal to the ellipse at the given point:
$\boxed{\text{SWAP}}$ $\boxed{\text{VAR}}$ (then $\boxed{\text{NXT}}$ or $\boxed{\leftarrow}$ $\boxed{\text{PREV}}$ as needed) **TNCO**.

Result (to 3 places): 2: Normal: 'y=1.187+.027*x'
                       1: Tangent: 'y=38.072-36.858*x'

Example: Find the equations of the lines normal and tangent to the parabola
$9y^2 - 64x + 54y - 26 = 0$, at the point where $x = 1$ and y is positive.

1. Compute the y-coordinate of the point of tangency: $\boxed{'}$ $\boxed{9}$ $\boxed{\times}$ $\boxed{\alpha}$ $\boxed{\leftarrow}$
$\boxed{Y}$ $\boxed{y^x}$ $\boxed{2}$ $\boxed{-}$ $\boxed{6}$ $\boxed{4}$ $\boxed{\times}$ $\boxed{\alpha}$ $\boxed{\leftarrow}$ $\boxed{X}$ $\boxed{+}$ $\boxed{5}$ $\boxed{4}$ $\boxed{\times}$ $\boxed{\alpha}$ $\boxed{\leftarrow}$ $\boxed{Y}$ $\boxed{-}$ $\boxed{2}$ $\boxed{6}$ $\boxed{\text{ENTER}}$ $\boxed{1}$
$\boxed{'}$ $\boxed{\alpha}$ $\boxed{\leftarrow}$ $\boxed{X}$ $\boxed{\text{STO}}$ $\boxed{'}$ $\boxed{\alpha}$ $\boxed{\leftarrow}$ $\boxed{Y}$ $\boxed{\text{ENTER}}$ $\boxed{1}$ $\boxed{0}$ $\boxed{\leftarrow}$ $\boxed{\text{SOLVE}}$ **ROOT** **ROOT**.
Result: 1.35889894354

2. Convert the previous result into a complex number representing the
point of tangency: $\boxed{1}$ $\boxed{\text{ENTER}}$ $\boxed{\text{SWAP}}$ $\boxed{\text{MTH}}$ $\boxed{\text{NXT}}$ **CMPL** **R→C**.

3. Enter the parabola as a list of coefficients: $\boxed{\leftarrow}$ $\boxed{\{\}}$ $\boxed{0}$ $\boxed{\text{SPC}}$ $\boxed{0}$ $\boxed{\text{SPC}}$ $\boxed{9}$
$\boxed{\text{SPC}}$ $\boxed{6}$ $\boxed{4}$ $\boxed{+/-}$ $\boxed{\text{SPC}}$ $\boxed{5}$ $\boxed{4}$ $\boxed{\text{SPC}}$ $\boxed{2}$ $\boxed{6}$ $\boxed{+/-}$ $\boxed{\text{ENTER}}$.

4. Compute the tangent and normal to the parabola at the given point: (SWAP)(VAR) (then (NXT) or (←)(PREV) as needed) **TNCON**.

> Result (to 4 places): 2: Normal: 'y=1.3433+.0156*x'
> 1: Tangent: 'y=65.3589-64*x'

Example: Find the equations of the lines normal and tangent to the hyperbola $16x^2 - 9y^2 - 64x + 54y - 26 = 0$, at the point where $x = 2$ and y is positive.

1. Compute the y-coordinate of the point of tangency: (')(1)(6)(×)(α)(←)(X)(yˣ)(2)(−)(9)(×)(α)(←)(Y)(yˣ)(2)(−)(6)(4)(×)(α)(←)(X)(+)(5)(4)(×)(α)(←)(Y)(−)(2)(6)(ENTER)(2)(')(α)(←)(X)(STO)(')(α)(←)(Y)(ENTER)(1)(0)(←)(SOLVE)**ROOT ROOT**. Result: 3.00000000001.

2. Convert the previous result into a complex number representing the point of tangency: (2)(ENTER)(SWAP)(MTH)(NXT)**CMPL R→C**.

3. Enter the hyperbola as a list of coefficients: (←)({ })(1)(6)(SPC)(0)(SPC)(9)(+/−)(SPC)(6)(4)(+/−)(SPC)(5)(4)(SPC)(2)(6)(+/−)(ENTER).

4. Compute the tangent and normal to the hyperbola at the given point: (SWAP)(VAR) (then (NXT) or (←)(PREV) as needed) **TNCON**.

> Result (to 3 places):
> 2: Normal: 'y=9.999E499*-9.999E499*x'
> 1: Tangent: 'y=3.000'

Note that the tangent is a horizontal line, so the normal is a vertical line (i.e. with an equation such as x = constant). The line reported by the TNCON program here is one method the HP 48 uses to report a vertical line as the result of a computation.

Translating and Rotating Conics

There are occasions where you may wish to rotate or translate a conic. The three programs, ROTCON, C→STD and TRNCON, make it easy to do this.

ROTCON (see page 299) takes a conic as a list of general-form coefficients from level 2 and an angle of rotation from level 1. Be sure to match the angle on level 1 with the current angle mode. ROTCON returns a list of coefficients for the transformed conic.

Example: Rotate the conic $9x^2 + 4y^2 + 36x - 8y + 4 = 0$ through an angle of 50°, and plot the result.

1. Enter the conic as a list of coefficients: ⤶ [{ }] [9] [SPC] [0] [SPC] [4] [SPC] [3] [6] [SPC] [8] [+/–] [SPC] [4] [ENTER].

2. Make sure that you're in degree mode (press ⤶ [RAD], if necessary) and then enter the angle: [5] [0] [ENTER].

3. Execute ROTCON: [VAR] (then [NXT] or ⤶ [PREV] as needed) █ROTCO█.

   <u>Result</u> (to 6 places): { .874787 -.710117 1 2.453375 -4.718681 }

4. Plot the rotated conic: [VAR] ([NXT] or ⤶ [PREV] as needed) █COMP█.

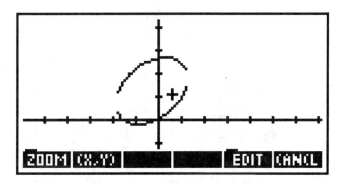

Example: Using the program C→STD (see page 280 for the listing), rotate the conic $9x^2 - 6xy - 4y^2 + 36x - 8y + 4 = 0$ to standard orientation, and plot the result.

1. Enter the conic as a list of coefficients: CANCEL ⬅️{ } 9 SPC 6 +/- SPC 4 +/- SPC 3 6 SPC 8 +/- SPC 4 ENTER.

2. Convert the conic to standard orientation: VAR (NXT or ⬅️PREV as needed) **C→ST**.

   Result: { 9.6589 0 -4.6589 36.8781 -.0909 4 }

3. Plot the rotated conic: VAR (NXT or ⬅️PREV as needed) **COMP**.

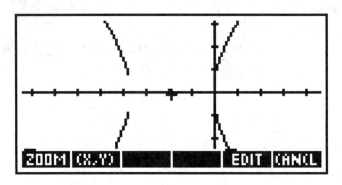

TRNCON (see page 316) takes a conic as a list of general-form coefficients from level 2 and a two-dimensional translation vector from level 1, returning a list of coefficients for the translated conic.

Example: Translate the conic $x^2 - 2y + 8x + 10 = 0$ along the vector [3 -4].

1. Enter the conic as a list of coefficients: CANCEL ⬅️{ } 1 SPC 0 SPC 0 SPC 8 SPC 2 +/- SPC 1 0 ENTER.

2. Enter the translation vector: ⬅️[ ] 3 SPC 4 +/- ENTER.

3. Translate the conic: VAR (then NXT or ⬅️PREV as needed) **TRNC**.

   Result: {1 0 0 2 -2 -13 }

4. Plot the translated conic: VAR (NXT or ⬅️PREV as needed) **COMP**.

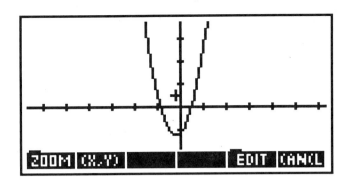

Example: Translate the conic $3x^2 + 3y^2 + 6x - 1 = 0$ so that it is centered on the origin.

1. Enter the conic as a list of coefficients and then make an extra copy:
   [CANCEL]⟨[ ]⟩[3][SPC][0][SPC][3][SPC][6][SPC][0][SPC][1][+/−][ENTER] [ENTER].

2. Note that the conic is a circle ($A = C$). Find the center: [VAR]([NXT] or ⟨[PREV] as needed) **G⇒CIR** [◀]. <u>Result</u>: (-1, 0)

3. Take the negative of the center and make it a translation vector. [+/−] [MTH][NXT] **CMPL** **C⇒R** [MTH] **VECTR** **⇒V2** . <u>Result</u>: [1 0] (Caution: If flag -19 is set, ⇒V2 gives a complex number.)

4. Translate the conic: [VAR] (then [NXT] or ⟨[PREV] as needed) **TRNC** .

   <u>Result</u>: {3 0 3 0 0 -4 }

4. Plot the translated conic: [VAR]([NXT] or ⟨[PREV] as needed) **CONP** .

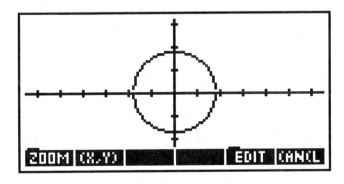

P. PROGRAM LISTINGS

On Using the Programs

This Appendix contains a listing of all of the programs referred to throughout this book. The programs are sorted alphabetically by name (numerals come after letters and special symbols are ignored in the sorting). Look at this example entry:

NAME   **Brief Program Description**   (page)

Program Size   Program Checksum

1: Stack Inputs	====>	1: Stack Outputs

« Program listing »

The page number refers you to the spot in the text where the program is first described. To compute the program size and checksum, with the program in memory, put its name (within single quotes) onto the stack and press ⟵ MEMORY **BYTES**.

To use a program, you must have it properly stored in a variable (its name) within the current directory path. (*Note:* If you're using an HP 48G, you will not be able to fit all of these programs into the 32K storage at once; you will need to pick and choose.) As with all HP 48 variables, you must be careful to avoid name conflicts with other variables in the current directory path. One suggestion: Put the programs into the one or more subdirectories and then create custom menus to assist you in organizing and accessing them. For details about custom menus, see your user's manual or Grapevine's *Easy Course in Programming the HP 48 G/GX*.

If you have a bit of programming aptitude, these programs can be modified to suit your tastes and/or needs. Most of them have not been rigorously groomed for error-trapping, speed, or memory efficiency; they are designed simply to work well with the examples in this course and with related work. Also, you may wish to modify the input or output of the programs. For example, geometric points may be expressed as either complex numbers or as two-element vectors, depending on the context in which you're working.

Whether you use these programs as is or otherwise, above all you should *practice* using them *before* needing them in an important situation. You must understand how they work, how fast are they, how to interpret their outputs, and the nature of their limitations (special cases of functions or flag settings).

Of course, each program should work flawlessly, but bugs (and typos) are, unfortunately, facts of life with software and other creative works. If you have a problem with a program, you may contact the publisher, but first, check again:

- *Have you correctly entered the program?* Use the checksum to verify this.

- *Are you correctly using the program?* Double-check the types and order of your inputs and the types and ranges of your graph settings. Also, note that some programs use ("call") other programs. Such instances are noted in these listings in *Boldface Italics*.

APOLY

Analyze a Polynomial (120)

500 bytes   #7AE3h

3:		3:	number of sign changes for p and –p
2:		2:	endpoints of range of real roots
1:	polynomial ====>	1:	polynomial

```
«  DUP SIZE 1 GET 0 ROT DUP
     → n s p q
     «  1 n
      FOR j q DUP j GET -1 n j - ^ * j SWAP PUT 'q' STO
      NEXT p q 2 →LIST 1
        «  OBJ→ 1 GET →LIST SIGN
          «  → a b
            «  IF a b + 0 == a 0 ≠ AND
              THEN s 1 + 's' STO
              END
              IF b
              THEN b
              ELSE a
              END
            »
          »  STREAM DROP s 0 's' STO
        »  DOLIST "Signs" →TAG p RROOTS 1
        «  IF DUP TYPE
          THEN DROP
          END
        »  DOLIST
          IF DUP SIZE 2 <
          THEN DUP 1 GET +
          END
          DUP « MIN » STREAM FLOOR 1 - SWAP « MAX »
          STREAM CEIL 1 + 2 →LIST "Range" →TAG p
      »
    »
»
```

A→Q

Rationalize an Array (143)

203.5 bytes #81FEh

| 1: | array | ====> | 1: | symbolic array with rational elements |

```
« RCLF -3 CF SWAP OBJ→ OBJ→
  IF 1 ==
  THEN 1 SWAP
  END 9 FIX
  → row col
  « 1 row
    FOR k 1 col
       START →Q col ROLLD
       NEXT col →LIST col row k - * k + ROLLD
    NEXT
    IF row 1 >
    THEN row →LIST
    END SWAP STOF
  »
»
```

CIR→G

Convert Circle Parameters to General Form (250)

97 bytes #4B7Eh

| 2: | (h,k) | | 2: | |
| 1: | radius | ====> | 1: | { A B C D E F } |

```
« → c r
  « 1 0 1 c -2 * C→R c C→R SQ SWAP SQ + r SQ - 6 →LIST 5
  FIX →Q STD
  »
»
```

CMPOS

Composite of Two Functions (22)

71 bytes #FADDh

3:	f		3:	
2:	g		2:	
1:	variable name	====>	1:	f o g

```
« → f g v
  « -3 CF g v STO f EVAL v PURGE
  »
»
```

COFACTR

Find Cofactor Matrix (154)

204 bytes #735Dh

1:	square matrix	====>	1:	cofactor matrix

```
« DUP DUP SIZE OBJ→ DROP
   → cof m r c
   « 1 r
     FOR i 1 c
        FOR j m i j 3 ROLLD ROW- DROP SWAP COL- DROP DET
           -1 i j + ^ * 1 →ARRY cof i j 2 →LIST ROT REPL
           'cof' STO
        NEXT
     NEXT cof
   »
»
```

COLIN?

Test for Collinearity of Point and Line (186)

168.5 bytes #DB81h

2:	direction vector of line		2:	
1:	point (in vector form)	====>	1:	1 if collinear; 0 if not

```
« DUP SIZE 1 GET
   → v p n
   « 1 n
     FOR k p k GET v k GET
        IF DUP 0 ==
        THEN DROP MINR
        END /
     NEXT n →LIST 2 « == » DOSUBS
     IF 0 POS
     THEN 0
     ELSE 1
     END
   »
»
```

CONPLT

Plot Conic From General Form (248)

1028 bytes #5AE6h

1:	{A B C D E F}	====>	1:	

```
« -3 CF { PICT PPAR } PURGE DUP OBJ→ DROP
   → con a b c d e f
   « IF b
     THEN b a c - / ATAN 2 /
     ELSE 0
     END
     → α
```

```
«  α COS SQ a * α COS α SIN b * * + α SIN SQ c * +
   α SIN SQ a * α SIN α COS b * * - α COS SQ c * +
   α COS d * α SIN e * + α COS e * α SIN d * -
 → ap cp dp ep
«   IF ap
   THEN dp NEG ap 2 * /
   ELSE ep SQ 4 cp f * * - 4 cp dp * * / dp NEG
        4 cp * / 20 * +
   END
   IF cp
   THEN ep NEG cp 2 * /
   ELSE dp SQ 4 ap f * * - 4 ap ep * * / ep NEG
        4 ap * / 20 * +
   END
   IF ap cp XOR
   THEN -22 SF ep ap / ABS dp cp / ABS -22 CF MIN
        5 *
   ELSE f NEG dp SQ ap 4 * / + ep SQ cp 4 * / +
        ap / ABS √
   END
 → h k r
«   con { 'x^2' 'x*y' 'y^2' x y 1 } * ΣLIST STEQ
    CONIC 'x' INDEP 'y' DEPND h r 3.6 * - DUP
    7.2 r * + XRNG k r 1.8 * - DUP 3.6 r * +
    YRNG DRAX DRAW PICTURE
  »
»
  »
   »
    »
»
```

CRAMER

Apply Cramer's Rule

(152)

249 bytes   #6D0Bh

2:	augmented matrix		2:	list of Cramer determinants
1:	list of variables	====>	1:	list of solutions

```
«  SWAP DUP SIZE 2 GET DUP ROT SWAP COL- SWAP DUP DET
 → v c b a d
«   IF d ABS .0000000001 <
   THEN "Ill-conditioned Matrix" DOERR
   ELSE 1 c 1 -
        FOR k a k COL- DROP b k COL+ DET
        NEXT c 1 - →LIST DUP d / v →TAG d ROT + SWAP
   END
  »
»
```

279

C→STD   Rotate Conic to Standard Orientation   (272)

320.5 bytes   #A291h

1: { A B C D E F }	====>	1: { A' 0 C' D' E' F' }

```
« OBJ→ DROP
  → a b c d e f
  « b a c -
    IF DUP
    THEN / ATAN 2 /
    ELSE DROP2 0
    END
    → θ
    « a θ COS SQ * b θ COS θ SIN * * + c θ SIN SQ * +
      0 a θ SIN SQ * b θ SIN θ COS * * - c θ COS SQ *
      + d θ COS * e θ SIN * + e θ COS * d θ SIN * - f
      6 →LIST
    »
  »
»
```

DMTRC   Create Dimetric Projection   (232)

326.5 bytes   # EE5h

1: factor (between 0 and 1)	====>	1: 4x4 matrix

```
« → f
  « DEG 'TAN(θ)^2+SIN(θ)^2-TAN(θ)^2*SIN(θ)^2=f^2' 'θ' 20
    ROOT
    → th
    « th SIN SQ DUP 1 SWAP - / √ ASIN
      → ph
      « ph COS th SIN ph SIN * ph SIN th COS * NEG 0
        0 th COS th SIN 0 ph SIN ph COS th SIN * NEG
        ph COS th COS * 0 0 0 0 1 { 4 4 } →ARRY
      »
    »
  »
»
```

DtoL   Find Distance from Point to a Line   (189)

83 bytes   #C733h

2: point (vector form) 1: line (array form)	====>	1: distance

```
« →ROW DROP OVER -
  → q p d
  « q p - d CROSS ABS d ABS /
  »
»
```

ELP→G   **Convert Ellipse Parameters to General Form**   (255)

195 bytes   #9141h

2:	(h, k)		2:	
1:	{ a b θ }	====>	1:	{ A B C D E F }

```
«  SWAP C→R ROT OBJ→ DROP
   →  h k a b θ
   «  b SQ θ a SQ h b SQ * 2 * NEG k a SQ * 2 * NEG h SQ
      b SQ * k SQ a SQ * + a SQ b SQ * - 6 →LIST θ ROTCON
   »
»
```

FINV   **Function Inverse**   (23)

168.5 bytes   #FC6Eh

2:	function		2:	
1:	variable name	====>	1:	modified array

```
«  RCLF
   →  f υ flags
   «  -3 CF υ PGALL 'τ↓' PGALL f 'τ↓' = υ ISOL υ 'τ↓'
      STO EVAL OBJ→ DROP2 SWAP DROP υ PURGE 'τ↓' PURGE
      flags STOF
   »
»
```

FMPLT   **Family Plot**   (10)

1185.5 bytes   #9DDFh

1:		====>	1:

```
«  RCLF → flgs
   «  {-3 -55 } CF FUNCTION .1 RES RAD
      IFERR RCEQ
      THEN { NOVAL }
      END PPAR DUP 3 GET
      IF DUP TYPE 5 ==
      THEN 1 GET
      END { NOVAL NOVAL } + ROT SWAP + SWAP DUP 1 GET RE
      SWAP 2 GET RE 2 →LIST +
      → flds
      «  WHILE "FAMILY PLOT" { { "EQ:" "ENTER THE FUNCTION"
            8 9 } { } { "INDEP:" "SPECIFY INDEPENDENT
            VARIABLE" 6 } { "VARY:" "SPECIFY THE VARIABLE TO
            CHANGE" 6 } { "VALS:" "SPECIFY THE VALUES TO
            USE" 5 } { } { "XMIN:" "ENTER MINIMUM HORIZONTAL
            VALUE" 0 } { "XMAX:" "ENTER MAXIMUM HORIZONTAL
            VALUE" 0 } } { 2 4 } flds DUP
            INFORM
         REPEAT DUP 'flds' STO REVLIST OBJ→ DROP DUP STEQ
```

```
     6 ROLLD INDEP OVER SIZE MAXR →NUM DUP NEG { }
   → j xmax xmin vals var lines ymin ymax q
   « xmin xmax XRNG vals 1 « →NUM » DOLIST SORT
     DUP SIZE 1 + 2 / SWAP DUP 3 PICK FLOOR
     GET SWAP ROT CEIL GET + 2 / 1.5 * var STO
     AUTO PPAR DUP 1 GET IM ymin MIN SWAP 2 GET
     IM ymax MAX YRNG CLLCD ERASE DRAX # 1h # 1h
     1 lines
     FOR line
         BLANK PICT { # 1h # 1h } ROT REPL
         var DUP vals line GET DUP 3 ROLLD
         SWAP STO PPAR 3 GET
         IF DUP TYPE 5 ==
         THEN 1 GET
         END
         PGALL RCEQ EVAL q SWAP + 'q' STO = PICT
         { # 1h # 1h } ROT 1 →GROB DUP SIZE
         5 ROLLD 5 ROLLD REPL DRAW
     NEXT
     q STEQ PICTURE DROP2 j STEQ
   »
   END
   »
   flgs STOF
  »
»
```

G→A   **Convert Line from General to Array Form**   (193)

184 bytes   **#D264h**

1:	line (general form)	====>	1: line (array form)

```
« { -3 -19 } CF
  → g
  « { x y } 1 « PGALL » DOLIST
    { 1 2 } DUP 1 « 'x' STO g 'y' 0 ROOT » DOLIST
    →V2 OBJ→ ROW→ { x y } PURGE
  »
»
```

G→CIR   **Find Circle Parameters from General Form**   (250)

215 bytes   **#9E69h**

2:			2: (h,k)
1:	{ A B C D E F }	====>	1: radius

```
« OBJ→ DROP
  → a b c d e f
  « IF a c == b 0 == AND
    THEN d NEG a 2 * / e NEG a 2 * / R→C d SQ e SQ +
       4 a f * * - 4 a SQ * / √
    ELSE "Not a circle" DOERR
```

```
        END
    »
»
```

G→ELP   **Find Ellipse Parameters from General Form**   (255)

217 bytes   #7850h

2:		2: (h,k)
1: {A B C D E F}	====>	1: {a b θ}

```
« DUP C→STD OBJ→ DROP
  → a b c d e f
  « d a 2 * / NEG e c 2 * / NEG R→C d SQ a 4 * / e SQ c
    4 * / + f - DUP a / √ SWAP c / √ 4 ROLL OBJ→ 4 DROPN
    ROT SWAP - / ATAN 2 / 3 →LIST
  »
»
```

G→HYP   **Find Hyperbola Parameters from General Form**   (261)

286 bytes   #1EF5h

2:		2: (h,k)
1: {A B C D E F}	====>	1: {a b θ}

```
« DUP C→STD OBJ→ DROP
  → con a b c d e f
  « d 2 a * / NEG e 2 c * / NEG R→C c d SQ * a e SQ * -
    4 a c f * * * - DUP 4 a SQ c * * / ABS √ SWAP 4 a c
    SQ * * / ABS √ DEG con 2 GET con 1 GET con 3 GET - /
    ATAN 2 / 3 →LIST
  »
»
```

G→PBL   **Find Parabola Parameters from General Form**   (260)

453 bytes   #6136h

2:		2: (h, k)
1: {A B C D E F}	====>	1: {p θ}

```
« DUP C→STD OBJ→ DROP
  → con a b c d e f
  « IF a
    THEN
        IF c
        THEN "Not a parabola"
        ELSE  d 2 a * / NEG d SQ 4 a f * * - 4 a e * * /
              R→C e 4 a * / NEG con 2 GET con 1 GET con 3
              GET - / ATAN 2 / 2 →LIST
        END
    ELSE
```

```
            IF c
            THEN  e SQ 4 c f * * - 4 c d * * / e 2 c * / NEG
                  R→C d 4 c * / NEG con 2 GET con 1 GET con 3
                  GET - / ATAN 2 / 2 →LIST
            ELSE "Not a parabola"
            END
        END
      »
    »
```

HYP→G   Convert Hyperbola Parameters from General Equation   (261)

195 bytes   #C3E6h

2:	(h,k)		2:	
1:	{ a b θ }	====>	1:	{ A B C D E F }

```
« SWAP C→R ROT OBJ→ DROP
  → h k a b θ
  «  b SQ 0 a SQ NEG 2 h b SQ * * NEG 2 k a SQ * * h SQ
     b SQ * a SQ k SQ * + a SQ b SQ * - 6 →LIST 0 ROTCON
  »
»
```

I→GEN   Convert Line from Slope-Intercept to General Form   (189)

201 bytes   #4BA8h

2:			2:	vector of coefficients
1:	line (slope-intercept form)	====>	1:	line (general form)

```
« { -2 -3 } CF
  → i
  « { 0 2 1 } 1 « 'x' STO i 'y' 0 ROOT » DOLIST OBJ→
    DROP - -1 ROT 3 →LIST DUP { x y 1 } * OBJ→ DROP + +
    COLCT 0 = SWAP OBJ→ →ARRY SWAP 8 FIX →Q STO { x y }
    PURGE
  »
»
```

INPLOT   Plot System of Inequalities   (161)

758 bytes   #D331h

1:		====>	1:	

```
« -3 CF RCEQ DUP SIZE
  IF DUP 2 <
  THEN DROP2 513 DOERR
  END { } { } PPAR 1 GET C→R PPAR 2 GET C→R OVER 5 PICK -
  PICT SIZE DROP B→R /
  → oldeqns numeqns neweqns pts xmin ymin xmax ymax step
  « oldeqns 1 « OBJ→ DROP2 = » DOLIST DUP 'neweqns' STO
```

```
            STEQ ERASE { # 0h # 0h } PVIEW DRAX DRAW oldeqns DUP
            STEQ 1 « OBJ→ DROP2 SWAP DROP » DOLIST 'neweqns' STO
            xmin xmax
            FOR x x ymax R→C DUP PIX? DUP2 « PIXOFF » « PIXON »
               IFTE x 'X' STO ymin ymax 2 →LIST neweqns 1 « →NUM
               » DOLIST + SORT 'pts' STO 1 numeqns 1 +
               FOR n pts n n 1 + SUB OBJ→ DROP DUP2 + 2 / 'Y'
                  STO
                  IF 1 1 numeqns
                    FOR m oldeqns m GET →NUM AND
                    NEXT
                  THEN x SWAP R→C x ROT R→C LINE
                  ELSE DROP2
                  END
               NEXT « PIXOFF » « PIXON » IFTE step
            STEP PICTURE
      »
    »
```

ISMTRC   **Create Isometric Projection**   (233)

207.5 bytes   #3C3Dh

1:	====>	1:

```
« 1 3 / √ ASIN 1 2 / √ ASIN
  → th ph
  « ph COS th SIN ph SIN * ph SIN th COS * NEG 0 0 th
    COS th SIN 0 ph SIN ph COS th SIN * NEG ph COS th
    COS * 0 0 0 0 1 { 4 4 } →ARRY
  »
»
```

LCON?   **Find Intersection of Line and Conic**   (265)

887 bytes   #4C68h

2: {A B C D E F}		2:
1: 'y=mx+b'	====>	1: { (intersection points) }

```
« → con line
  « { x y } 1 « PGALL » DOLIST con { 'x^2' 'x*y' 'y^2'
    x y 1 } * ∑LIST DUP -3 CF line DEFINE EVAL
    IF con DUP 2 GET SQ 4 con 1 GET con 3 GET * * - DUP
    THEN
        IF 0 <
        THEN
            IF con C→STD DUP 1 GET SWAP 3 GET ==
            THEN G→CIR
            ELSE G→ELP 1 GET
            END
        ELSE G→HYP DUP 1 GET SQ SWAP 2 GET SQ + √ 10 *
        END DUP ROT C→R DROP DUP ROT - 3 ROLLD +
    ELSE DROP G→PBL DROP C→R DROP DUP 10 - SWAP 10 +
    END
```

```
    →  scon sol s a
    «  0
       DO sol 'x' s ROOT sol 'x' a ROOT
          IF DUP2 ==
          THEN DROP2 s 10 - 's' STO a 10 + 'a' STO 1 + 1
               SF
          ELSE
               IF DUP2 - ABS .0001 ≤
               THEN + 2 / 1 →LIST
               ELSE 2 →LIST
               END
               1 « DUP line SWAP 'x' STO 'y' 0 ROOT R→C »
               DOLIST 1 CF
          END
       UNTIL DUP TYPE 5 == OVER 3 == OR
       END
       IF scon EVAL 4 RND
       THEN DROP { }
       ELSE
            IF 1 FS?C
            THEN x 1 →LIST
            END
       END SWAP DROP { x y } PURGE
    »
  »
»
```

LINPRG

Linear Programming

2009 bytes   #13CFh

3:		3:	list of basic variables
2:		2:	final tableau
1:	====>	1:	list of tagged solutions or message string

```
« 0 'MARKER' STO DEPTH 'depth' STO RCLF 'flags' STO -3 CF
  "LINEAR PROGRAMMING" { { "OBJECTIVE:" "ENTER OBJECTIVE
  FUNCTION" 9 } { "CONSTRAINTS:" "ENTER LIST OF ALG.
  CONSTRAINTS" 5 } { "VARS:" "ENTER LIST OF INDEP.
  VARIABLES" 5 } { "MAX OR MIN?" "COMPUTE MAX OR MIN OF
  OBJECTIVE?" 2 } } { 1 2 } { NOVAL NOVAL NOVAL "MAX" }
  { NOVAL NOVAL NOVAL "MAX" }
  INFORM
  IF
  THEN CLLCD "Solving . . ." 3 DISP 5 CF OBJ→ DROP
       IF "MIN" SAME
       THEN 5 SF
       END DUP SIZE ROT DUP SIZE { } DUP DUP DUP 'last' STO
       'nvars' STO 'bvars' STO
       → of vars n constr m eqns
       « 1 n
         FOR i nvars i + 'nvars' STO
         NEXT 1 m
         FOR k bvars k n + + 'bvars' STO constr k GET OBJ→
             { < ≤ = ≥ > } SWAP POS
             IF DUP { 4 5 } SWAP POS
```

P. PROGRAM LISTINGS

```
                THEN DROP2 = -1 * eqns SWAP + 'eqns' STO
                ELSE
                    IF { 1 2 } SWAP POS
                    THEN DROP
                    ELSE DROP bvars k 0 PUT 'bvars' STO
                    END = eqns SWAP + 'eqns' STO
                END
            NEXT eqns of
            IF 5 FS?
            THEN -1 *
            END + 'eqns' STO 1 n
            FOR k 0 vars k GET STO
            NEXT eqns OBJ→ 1 SWAP
            START m 1 + ROLL →NUM
            NEXT m 1 + →ARRY NEG
        → b
    «   1 n
            FOR k 1 vars k GET STO eqns OBJ→ 1 SWAP
                FOR j m 1 + ROLL →NUM b j GET +
                NEXT 0 vars k GET STO
            NEXT n m 1 + 2 →LIST →ARRY TRN b n 1 + COL+
            'a' STO
            DO m n bvars nvars a PHASE1 m n bvars nvars a 1
                SIMPLEX
            UNTIL a DUP SIZE 2 GET COL- SWAP DROP OBJ→ 1
                GET →LIST 1 bvars SIZE SUB SIGN -1 POS 0 ==
                1 FS? OR
            END
            IF 1 FS?C
            THEN 5 DROPN bvars a "No feasible solution
                exists" DUP MSGBOX
            ELSE
                IF a m 1 + ROW- SWAP DROP OBJ→ 1 GET →LIST 1
                    nvars SIZE SUB 0 POS
                THEN 5 ROLLD 4 DROPN bvars SWAP "Solution is
                    unbounded" DUP MSGBOX
                ELSE 5 DROPN a nvars SIZE 1 + COL- SWAP DROP
                    'solns' STO 1 vars SIZE
                    FOR k
                        IF nvars k GET DUP vars SIZE ≤ SWAP 0
                            > AND
                        THEN 0 vars nvars k GET GET STO
                        END
                    NEXT 1 bvars SIZE
                    FOR k
                        IF bvars k GET DUP vars SIZE ≤ SWAP 0
                            > AND
                        THEN solns k GET vars bvars k GET GET
                            STO
                        END
                    NEXT DEPTH depth - DROPN bvars a vars DUP
                    EVAL n →LIST SWAP →TAG "Solution found"
                    MSGBOX
                END
            END
        »
    »
END flags STOF VARS DUP 'MARKER' POS 1 SWAP SUB PURGE
»
```

LIN2?
Determine Relationship of 2 Lines
(200)

549 bytes   #3FD2h

2:	line 1 (array form)		2:	pt. of intersection or "relationship"
1:	line 2 (array form)	====>	1:	1 if intersect; 0 if not

```
«  →ROW DROP OVER - ROT →ROW DROP OVER - DUP SIZE 1 GET
   → p1 d1 p2 d2 n
   «  IF d1 d2 DOT ABS d1 ABS d2 ABS * / 10 RND 1 ==
      THEN
         IF d2 p1 COLIN?
         THEN "Concurrent" 0
         ELSE "Parallel" 0
         END
      ELSE p1 p2 - d2 d1 -
         → u v
         «  1 n
            FOR k u k GET v k GET
               IF DUP 0 ==
               THEN DROP MINR
               END / 10 RND
               IF DUP 0 ==
               THEN DROP n 1 - 'm' STO
               END
            NEXT m →LIST DUP
            IF m 1 >
            THEN 2 « == » DOSUBS
            END
            IF 0 POS n 2 > AND
            THEN DROP "Skew" 0
            ELSE 1 GET d1 * p1 + 1
            END
         »
         END 'm' PURGE
      »
   »
```

LPL→P
Find Intersection Point of Line and Plane
(211)

198.5 bytes   #780Bh

3:	position vector of line		3:	
2:	direction vector of line		2:	
1:	vector of plane's coefficients	====>	1:	pt. of intersection or "relationship"

```
«  4 COL-
   → p d n nd
   «  IF n d DOT 0 ==
      THEN
         IF p n DOT NEG nd ==
         THEN "Coplanar"
         ELSE "Parallel"
         END
      ELSE nd NEG n p DOT - n d DOT / d * p +
      END
```

```
    »
    »
```

LTRIM   **Trim Zeroes from Left of Array**   (289)

135.5 bytes   #200Ah

1: array	====>	1: trimmed array

```
«  DUP RNRM
   IF .00001 <
   THEN DROP 0 1 →ARRY
   ELSE OBJ→ 1 GET 1 +
       WHILE DUP ROLL DUP ABS .00001 <
       REPEAT DROP 1 -
       END OVER ROLLD 1 - →ARRY
   END
»
```

NRMLZ   **Normalize Object Array after Transformation**   (234)

119.5 bytes   #4D6Dh

1: object array	====>	1: normalized object array

```
«  DUP SIZE OBJ→ DROP
   → a m n
   «  1 m
      FOR i a DUP { i n } GET INV i RCI 'a' STO
      NEXT a
   »
»
```

PADD   **Polynomial Addition**   (109)

172.5 bytes   #64CBh

2: polynomial 1		2:
1: polynomial 2	====>	1: P1 + P2

```
«  RCLF 3 ROLLD -55 CF
   IFERR +
   THEN OVER SIZE 1 GET OVER SIZE 1 GET -
       IF DUP 0 <
       THEN ABS ROT SWAP
       END
       → a d
       «  1 d
          START 0
          NEXT a OBJ→ 1 GET d + →ARRY +
       »
   END
   LTRIM SWAP STOF
»
```

PAR→I   Convert Line (2D) from Parametric to Slope-Intercept Form   (193)

167 bytes   **#35F5h**

1:	list of parametric eqns of line (2D) ====>	1:	line (slope-intercept form)

```
« RCLF -3 CF SWAP
  IF DUP SIZE 2 ≠
  THEN DROP 515 DOERR ELSE 9 FIX 1
       « 't' ISOL COLCT OBJ→ DROP2 SWAP DROP EXPAN EXPAN
         COLCT
       »
       DOLIST OBJ→ DROP = 'y' ISOL EXPAN EXPAN COLCT →Q
  END SWAP STOF
»
```

PBL→G   Convert Parabola Parameters to General Form   (260)

134 bytes   **#D7B9h**

2:	(h,k)		2:	
1:	{ p θ }	====>	1:	{ A B C D E F }

```
« SWAP C→R ROT OBJ→ DROP
  → h k p θ
  « 0 0 1 p 4 * NEG k 2 * NEG k SQ 4 p h * * + 6 →LIST
    θ ROTCON
  »
»
```

PCONV   Convert to Polynomial Form   (133)

680 bytes   **#4D20h**

2:	array		2:	
1:	program	====>	1:	modified array

```
« -3 CF { [ 0 ] [ 1 ] } { N D } STO →RPN DUP SIZE
  « → n d
    « N d PMULT D n PMULT 'OP' EVAL 'N' STO D d PMULT
      'D' STO
    »
  » → p n ←pdiv
  « 1 n
    FOR k 'p' k GET
        IF DUP TYPE
        THEN
            IF DUP TYPE { 6 7 } SWAP POS
            THEN DROP [ 1 0 ] 1 →LIST
            ELSE { + - * ^ NEG DEC } SWAP POS
                 IF DUP
                 THEN { PADD PSUB PMULT PPOWER PSUB
                   « EVAL DUP DROP » } SWAP GET 1 →LIST
                 ELSE DROP p k 1 + GET { + - } SWAP POS
```

```
                              { PADD PSUB } SWAP GET 'OP' STO
                              { ←pdiv DEC }
                      END
                  END
               ELSE 1 →ARRY 1 →LIST
               END p k ROT REPL 'p' STO
          NEXT p EVAL D PMULT N PADD D { OP N D } PURGE
    »
  »
```

PDIVIDE   **Polynomial Division**   (112)

320.5 bytes   #F513h

4:			4:	quotient array
3:			3:	numerator of remainder
2:	polynomial 1		2:	denominator of remainder
1:	polynomial 2	====>	1:	symbolic result

```
«  LTRIM DUP OBJ→ OBJ→ DROP →LIST ROT LTRIM OBJ→ OBJ→ DROP
   →LIST SWAP DUP2 SIZE SWAP SIZE
   IF OVER - DUP 0 <
   THEN DROP2 { 0 } 3 ROLLD
   ELSE SWAP ROT DUP 1 GET
      → n p2 t
        « { } 3 ROLLD 0 SWAP
          START DUP 1 GET t / ROT OVER + 3 ROLLD 1 n
             FOR d
                OVER d GET p2 d GET 3 PICK * - ROT d
                ROT PUT SWAP
             NEXT DROP 2 OVER SIZE MIN 1E499 SUB
          NEXT
        »
      SWAP OBJ→ →ARRY SWAP OBJ→ →ARRY ROT REMNDR
   END
  »
```

PD→P   **Convert Line from Position-Direction to Parametric Form**   (192)

184.5 bytes   # 269h

2:	position vector of line		2:	
1:	direction vector of line	====>	1:	list of parametric eqns of line

```
«  -3 CF DUP SIZE 1 GET
   → p v n
   « IF n 2 >
     THEN { x y z }
     ELSE { x y }
     END v OBJ→ 1 GET →LIST 1 n
     START 't'
     NEXT n →LIST * p OBJ→ 1 GET →LIST ADD = 9 FIX →Q STD
   »
  »
```

PERSP

Create Perspective Projection (238)

453 bytes #4FBDh

3:	object array		3:	
2:	translation vector		2:	
1:	eyepoint vector	====>	1:	transformed object array

```
« DUP ABS
  → a t c d
  « c 1 GET d / ASIN c 2 GET NEG d / ASIN c 3 GET NEG t
    OBJ→ DROP
    → θ f k l m n
    « a θ COS θ SIN f SIN * θ θ SIN f COS * k / NEG θ f
      COS θ f SIN k / θ SIN θ COS f SIN * NEG θ θ COS f
      COS * k / θ COS l * θ SIN n * + f COS m * θ SIN
      l * θ COS n * - f SIN * + θ f SIN m * θ SIN l *
      + θ COS n * + k / 1 + { 4 4 } →ARRY * NRMLZ
    »
  »
»
```

PGALL

Purge Variable in Path (281)

92 bytes #68E4h

1:	variable name	====>	1:	

```
« PATH
  → name path
  « DO name name PURGE EVAL UPDIR
    UNTIL TYPE 6 ==
    END path EVAL
  »
»
```

PHASE1

Perform Phase 1 Adjustments on LP Tableau (164)

1315 bytes #3F86h

5:	number of constraints		5:	
4:	number of decision variables		4:	
3:	list of indexes of basic variables		3:	
2:	list of indexes of non-basic variables		2:	
1:	tableau	===>	1:	

```
« 'a' STO 'nvars' STO 'bvars' STO
  → m n
  « WHILE bvars 0 POS DUP
    REPEAT DUP 'r' STO a SWAP ROW- SWAP DROP OBJ→ 1 GET
      →LIST 1 nvars SIZE SUB DUP
      IF « MAX » STREAM DUP ABS .0001 >
      THEN POS 's' STO
```

```
          ELSE DROP
             IF DUP « MIN » STREAM DUP ABS .0001 >
             THEN POS 's' STO
             ELSE DROP
             END
          END m n bvars nvars a r s PIVOT 5 DROPN
END DROP a DUP SIZE 2 GET COL- SWAP DROP OBJ→ 1 GET
→LIST DUP SIGN
→ bl signs
«   IF signs 1 bvars SIZE SUB -1 POS
    THEN 1 m
       FOR k signs k GET
          IF 0 ≥
          THEN signs k 0 PUT 'signs' STO
          END
       NEXT a signs nvars SIZE 1 + 's' STO bvars SIZE
       1 + bvars SIZE 2 +
       FOR k k { 0 } REPL
       NEXT OBJ→ DUP a SIZE 1 GET
       IF ≠
       THEN SWAP DROP 1 -
       END →ARRY s COL+ 'a' STO nvars 0 + 'nvars' STO
       bl OBJ→
       IF DUP m - 1 -
       THEN ROT DROP SWAP DROP 2 -
       ELSE SWAP DROP 1 -
       END →LIST DUP « MIN » STREAM POS 'r' STO m n
       bvars nvars a r s PIVOT 5 DROPN 1 a SIZE 2 GET
       FOR j 0
       NEXT a SIZE 2 GET →ARRY
       → z
       «  a
          IF DUP SIZE 1 GET m 2 + ==
          THEN m 2 + 1 2 →LIST z REPL
          ELSE z m 2 + ROW+
          END 'a' STO 1 m
          FOR i
             IF a i nvars SIZE 2 →LIST GET SIGN -1 ==
             THEN a i ROW- z + SWAP DROP 'z' STO a m 2
                + 1 2 →LIST z REPL 'a' STO
             END
          NEXT m n bvars nvars a 2 SIMPLEX
          IF 1 FC? a DUP SIZE GET 0 ≤ AND
          THEN PHASE1
          ELSE 1 SF 5 DROPN
          END
       »
    END
»
»
»
```

293

PIVOT

Pivot LP Tableau on Given Element (164)

695.5 bytes #139Ch

7: number of constraints		7:
6: number of decision variables		6:
5: list of indexes of basic variables		5: number of constraints
4: list of indexes of non-basic variables		4: number of decision variables
3: tableau		3: list of indexes of basic variables
2: pivot row		2: list of indexes of non-basic variables
1: pivot column	===>	1: tableau

```
« 5 ROLLD 5 ROLLD 'a' STO 'nvars' STO 'bvars' STO
   → m n r s
   « a DUP r s 2 →LIST DUP 3 ROLLD GET INV DUP 4 ROLLD PUT
     s COL- SWAP ROT r RCI r ROW- ROT r ROW-
     → ap rp sp ars
     « 1 ap SIZE 1 GET
       FOR k ap k ROW- rp sp k GET * - k ROW+ 'ap' STO
       NEXT sp ars NEG * 'sp' STO ap rp r ROW+ sp ars r
       ROW+ s COL+ OBJ→ DUP OBJ→ DROP * SWAP OVER 2 +
       ROLLD →LIST 9 RND OBJ→ 1 + ROLL →ARRY 'a' STO bvars
       nvars s GET nvars bvars r GET s SWAP PUT 'nvars'
       STO r SWAP PUT 'bvars' STO
       WHILE nvars DUP 0 POS DUP
       REPEAT DUP a SWAP COL- DROP 'a' STO SWAP OBJ→ DUP 2
          + ROLL OVER 2 + SWAP - ROLL DROP 1 - →LIST
          'nvars' STO
       END DROP2 m n bvars nvars a
     »
   »
»
```

PL2→L

Find Intersection of Two Planes (206)

140 bytes #B902h

2: plane 1 (vector form)		2:
1: plane 2 (vector form)	====>	1: line of intersection (parametric form)

```
« DUP2 2 ROW→
   → p q a
   « a 4 COL- SWAP DUP →ROW DROP CROSS SWAP 3 COL- DROP
     ROT SWAP / OBJ→ DROP 0 3 →LIST 10 RND OBJ→ →ARRY SWAP
     PD→P
   »
»
```

PMULT   **Polynomial Multiplication**   (111)

202.5 bytes   #44DDh

		2:
2: polynomial 1		
1: polynomial 2	====>	1: P1 * P2

```
« DUP SIZE 1 GET
  → a na
    « DUP DUP 0 CON DUP SIZE 1 GET DUP na + 1 - 1 →LIST
      → b c nb nab
        « { 1 } nb + RDM 1 nb
          START a OBJ→ DROP c OBJ→ DROP
          NEXT
          nb DROPN nb nab + →ARRY * nab RDM
        »
    »
»
```

P→PD   **Convert Line from Parametric to Position-Direction Form**   (193)

378.5 bytes   #A688h

2:		2: position vector for line
1: list of parametric eqns of line	====>	1: direction vector for line

```
« { -3 -19 } CF 1
  « →RPN
    → c
    « c SIZE 'n' STO c c 2 c 't' POS DUP
      IF DUP 1 + c SWAP GET { NEG } 1 GET ==
      THEN 1 + -1
      ELSE 1
      END 't' STO 's' STO SUB
      IF DUP SIZE 2 <
      THEN 1 GET
      ELSE πLIST
      END EVAL { 0 } ROT s 1 + n 1 - SUB
      IF DUP 1 GET TYPE 18 ==
      THEN TAIL
      END + EVAL 2 →LIST
    »
  »
  DOLIST
  IF OBJ→ 2 >
  THEN →V3
  ELSE →V2
  END OBJ→ DROP SWAP { s n t } PURGE
»
```

295

PPOWER   **Raise a Polynomial to a Power**   (114)

175 bytes   #8CFAh

2:	polynomial		2:	
1:	power (array)	====>	1:	modified power

```
« 1 GET
   → P n
   «  1 DUP →ARRY
      WHILE n 0 >
      REPEAT
         IF n 2 MOD
         THEN P PMULT
         END n 2 / FLOOR 'n' STO
         IF n
         THEN P DUP PMULT 'P' STO
         END
      END
   »
»
```

PSUB   **Polynomial Subtraction**   (110)

28.5 bytes   #DAD9h

2:	polynomial 1		2:	
1:	polynomial 2	====>	1:	P1–P2

```
« NEG PADD »
```

P→SYM   **Polynomial to Symbolic**   (110)

115 bytes   #30BEh

2:	polynomial (array)		2:	
1:	polynomial variable	====>	1:	polynomial (symbolic)

```
« -3 CF
   → v
   «  OBJ→ OBJ→ DROP 0 SWAP 1
      FOR n
         n 1 + ROLL v n 1 - ^ * + -1
      STEP 10 FIX →Q STD
   »
»
```

P2→L   **Find Line Containing Two Points**   (182)

130 bytes   #B45Ch

2:	point 1 (2D-vector)		2:	
1:	point 2 (2D-vector)	====>	1:	slope-intercept equation of line

```
« -3 CF
  → p1 p2
  « 'y' p2 p1 - OBJ→ DROP SWAP / DUP 'x' * SWAP p1 OBJ→
    DROP SWAP ROT * NEG + + = 8 FIX →Q STD
  »
»
```

P2→PB   **Find Perpendicular Bisector from Two Points**   (184)

207 bytes   #AAE4h

2:	point 1 (2D-vector)		2:	
1:	point 2 (2D-vector)	====>	1:	equation of perp. bisector

```
« -3 CF
  → p1 p2
  « 'y' p2 p1 - OBJ→ DROP SWAP / NEG
    IF DUP
    THEN INV DUP 'x' * SWAP p1 p2 + 2 / OBJ→ DROP SWAP
         ROT * NEG + +
    ELSE DROP2 'x' p1 p2 + 2 / 1 GET
    END = 8 FIX →Q STD
  »
»
```

QDSOLV   **Solve Quadratic**   (121)

398 bytes   #915Dh

1:	quadratic (array)	====>	1:	list of solutions

```
« { -1 -3 } CF OBJ→ DROP
  → a b c
  « b SQ 4 a c * * - a 2 *
    → d e
    « IF d 0 ≤
      THEN
        IF d DUP IP ==
        THEN d ABS
          IF √ DUP IP ≠
          THEN d ABS 1 →LIST 'sqrt' APPLY i *
          ELSE d ABS √ i *
          END
        ELSE d ABS √ i *
        END
      ELSE
        IF d √ DUP IP ≠
```

```
        THEN d 1 →LIST 'sqrt' APPLY
        ELSE d √
        END
    END e / DUP b NEG e / DUP ROT + 3 ROLLD SWAP - 2
    →LIST -27 SF COLCT
  »
 »
»
```

RCOEF   Compute Symbolic Polynomial from Roots   (132)

239.5 bytes   **#9E93h**

1:	list of roots	====>	1: symbolic polynomial

```
« -3 CF 1 « EVAL » DOLIST OBJ→ →ARRY PCOEF
DUP OBJ→ 1 GET →LIST 1
  « IF DUP TYPE
    THEN DROP
    ELSE FP ABS
        IF DUP
        THEN 7 FIX →Q STD OBJ→ DROP2 SWAP DROP
        ELSE DROP 1
        END
    END
  » DOLIST
  « DUP2
    WHILE DUP
    REPEAT SWAP OVER MOD
    END DROP / * EVAL
  » STREAM * 'X' P→SYM
»
```

REMNDR   Compute Symbolic Remainder   (291)

276 bytes   **#EE73h**

3:	quotient array		3:
2:	numerator array		2:
1:	denominator array	====>	1: symbolic remainder

```
« -3 CF
  → q n d
  « IF n [ 0 ] ≠
    THEN q OBJ→ 1 GET
      → t k
      « 0 k →ARRY t 1 →ARRY d PMULT n PADD 'x' P→SYM
        d 'x' P→SYM / SWAP 'x' P→SYM SWAP +
      »
    ELSE q 'x' P→SYM
    END q n d 4 ROLL
  »
»
```

RFLCT
Create Reflection Transformation (227)

273.5 bytes #CA74h

2:	object array		2:	
1:	plane of reflection (vector form)	====>	1:	reflected object array

```
«  → a v
   «  IF v SIZE 1 GET 3 ==
      THEN v DUP 3 GET SWAP 3 0 PUT
      ELSE v DUP 4 GET SWAP 4 COL- DROP
      END
      → d n
      « 1 a SIZE 1 GET
         FOR i d NEG n a i ROW- SWAP 4 ROLLD DOT - n n DOT
            / n * 2 * a i ROW- SWAP DROP + i ROW+ 'a' STO
         NEXT a
      »
   »
»
```

ROTCON
Rotate Conic (271)

354.5 bytes #C3C2h

1:	{ A B C D E F }	====>	1:	{ A' B' C' D' E' F' }

```
«  SWAP OBJ→ DROP
   → θ a b c d e f
   «  a θ COS SQ * b θ COS θ SIN * * + c θ SIN SQ * + b θ
      COS SQ θ SIN SQ - * 2 c a - * + θ SIN θ COS * * a θ
      SIN SQ * b θ SIN θ COS * * - c θ COS SQ * + d θ COS
      * e θ SIN + e θ COS * d θ SIN * - f 6 →LIST DUP
      DUP 1 GET ABS SWAP 3 GET ABS MAX /
   »
»
```

ROT2D
Create 2D-Rotation Transformation Matrix (222)

174.5 bytes #C635h

2:	point at center of rotation (vector)		2:	
1:	angle of rotation	====>	1:	transformation matrix

```
«  SWAP OBJ→ DROP
   → x l m
   «  x COS x SIN 0 x SIN NEG x COS 0 l NEG x COS 1 - * m
      x SIN * + l NEG x SIN * m x COS 1 - * - 1 { 3 3 }
      →ARRY
   »
»
```

ROT3D — Create 3D-Rotation Transformation Matrix (225)

367.5 bytes   #AF7Bh

3:	position vector for axis of rotation		3:	
2:	direction vector for axis of rotation		2:	
1:	angle of rotation	====>	1:	transformation matrix

```
« 4 IDN
  → p d x i
  « d 0 4 COL+ d OBJ→ DROP
    → n a b c
    « 1 3
      FOR k n n k GET *
      NEXT [ 0 0 0 1 ] 4 ROW→ 1 x COS - * i x COS * +
      0 c b NEG 0 c NEG 0 a 0 b a NEG 0 0 0 0 0 { 4
      4 } →ARRY x SIN * +
      → r
      « i { 4 1 } p NEG REPL r * i { 4 1 } p REPL *
      »
    »
  »
»
```

→RPN — Convert Algebraic to RPN List (290)

189.5 bytes   #6FB6h

	1:	algebraic	====>	1:	list

```
« OBJ→
  IF OVER
  THEN → n f
    « 1 n
      FOR i
          IF DUP TYPE 9 SAME
          THEN →RPN
          END n ROLLD
      NEXT
      IF DUP TYPE 5 ≠
      THEN 1 →LIST
      END
      IF n 1 >
      THEN 2 n
          START +
          NEXT
      END f +
    »
  ELSE 1 →LIST SWAP DROP
  END
»
```

RROOTS   **Find Real Roots of Polynomial**   (130)

141 bytes   **#54ECh**

1: polynomial (array)	====>	1: list of real roots

```
« PROOT DUP SIZE 1 GET
   → r n
     « { } 1 n
       FOR k
          r k GET IM
          IF
          THEN r k GET +
          ELSE r k GET RE SWAP +
          END
       NEXT
     »
»
```

SCFACTR   **Find Symbolic Cofactor Matrix**   (159)

234.5 bytes   **#5DFCh**

1: square symbolic matrix	====>	1: symbolic cofactor matrix

```
« -3 CF DUP DUP 1 GET SIZE SWAP DUP SIZE
   → cof c m r
   « 1 r
     FOR i 1 c
        FOR j m i j 3 ROLLD SXROW DROP SWAP SXCOL DROP
           SDET -1 i j + ^ * cof DUP i GET j 4 ROLL PUT i
           SWAP PUT 'cof' STO
        NEXT
     NEXT cof
   »
»
```

SCOF   **Find Symbolic Cofactor**   (158)

254.5 bytes   **#EAE0h**

3: symbolic matrix		
2: row	2:	
1: column	====>	1: cofactor

```
« -3 CF 3 PICK DUP SIZE SWAP 1 GET SIZE DROP
   IF 1 ==
   THEN 3 DROPN 1
   ELSE → r c
     « OBJ→ OVER SIZE OVER 1 -
       → m n
       « r - 1 + ROLL DROP 1 n
         START n ROLL
            IF c 1 -
```

```
                    THEN DUP 1 ⊏ 1 - SUB SWAP ⊏ 1 + m SUB +
                    ELSE 2 m SUB
                    END
              NEXT n →LIST
          »
        » SDET
    END
»
```

SCRAMER
Apply Cramer's Rule to Symbolic Matrix
(159)

214.5 bytes   #BA68h

2:	symbolic matrix		2:	list of Cramer determinants
1:	list of variables	====>	1:	list of solutions

```
« -3 CF SWAP DUP 1 GET SIZE DUP ROT SWAP SXCOL
  SWAP DUP SDET
  → v c b a d
  « 1 c 1 -
    FOR k a k SXCOL DROP b k SNCOL SDET
    NEXT c 1 - →LIST DUP d / COLCT →Q v →TAG d ROT +
    COLCT →Q SWAP
  »
»
```

SCSWP
Swap Columns in Symbolic Array
(159)

77.5 bytes   #9B31h

3:	symbolic array			
2:	column 1		2:	
1:	column 2	====>	1:	modified array

```
« → s i j
  « s STRN i j SRSWP STRN
  »
»
```

SDET
Symbolic Determinant
(158)

136.5 bytes   #575Dh

1:	symbolic matrix	====>	1:	determinant

```
« -3 CF DUP DUP SIZE SWAP 1 GET SIZE DROP
  → a n
  « 0 1 n
    FOR i a i GET 1 GET a i 1 SCOF * -1 i 1 + ^ * +
    NEXT
  »
»
```

SDIV

Synthetic Division (stack version)

174.5 bytes   #4842h

2: array		2:
1: program	====>	1: modified array

```
« → p f
   « p p SIZE 1 GET 0
      → n s
      « 1 n
        FOR k
           p k GET s + DUP f * 's' STO
        NEXT
        → r
           « n 1 - →ARRY f SWAP r
           »
      »
   »
   's' PURGE
»
```

SIMPLEX

Apply Simplex Algorithm to LP Tableau

1166 bytes   #C97Fh

6: number of constraints		6:
5: number of decision variables		5: number of constraints
4: list of indexes of basic variables		4: number of decision variables
3: list of indexes of non-basic variables		3: list of indexes of basic variables
2: tableau		2: list of indexes of non-basic variables
1: 1 if Phase 2, or 2 if Phase 1	===>	1: tableau

```
« 4 ROLLD 'a' STO 'nvars' STO 'bvars' STO
   → m n t
   « { 3 4 6 } CF
     WHILE a m t + ROW- SWAP DROP OBJ→ 1 GET →LIST 1 nvars
        SIZE SUB DUP 'c' STO SIGN 1 POS 4 FS?C OR
     REPEAT a DUP SIZE 2 GET COL- SWAP DROP OBJ→ 1 GET
        →LIST 1 bvars SIZE SUB DUP 'b' STO
     IF 0 POS
     THEN 6 SF
     END
     IF 6 FS?
     THEN { } 1 nvars SIZE
        FOR k
           IF c k GET .0001 >
           THEN nvars k GET
              IF DUP
              THEN +
              ELSE DROP
              END
           END
        NEXT
     IF DUP { } SAME
     THEN DROP { 1 4 } SF
```

```
                    ELSE
                       IF DUP SIZE 1 ==
                       THEN 1 GET
                       ELSE « MIN » STREAM
                       END nvars SWAP POS 's' STO
                    END
                 ELSE c DUP « MAX » STREAM POS 's' STO
                 END { } 1 m
                 FOR i
                    IF a i s 2 →LIST GET .0001 >
                    THEN bvars i GET +
                    END
                 NEXT 'scol' STO
                 IF scol { } SAME
                 THEN { 1 4 } SF
                 ELSE
                    IF 6 FS?C
                    THEN scol
                       IF DUP SIZE 1 ==
                       THEN 1 GET
                       ELSE « MIN » STREAM
                       END bvars SWAP POS 'r' STO
                    ELSE 1 scol SIZE
                       FOR j
                          IF a bvars scol j GET POS s 2 →LIST GET
                             DUP .0001 >
                          THEN b bvars scol j GET POS GET SWAP ∕
                          END
                       NEXT scol SIZE →LIST DUP
                       IF DUP SIZE 1 ==
                       THEN 1 GET
                       ELSE « MIN » STREAM
                       END POS scol SWAP GET bvars SWAP POS 'r' STO
                    END
                 END m n bvars nvars a r s PIVOT 5 DROPN
              END m n bvars nvars a
       »
    »
```

SM→   Disassemble Symbolic Array   (158)

124 bytes   #46AAh

mn + 2:		mn + 2:	
...			... elements
3:		3:	
2:		2:	# of rows
1:	symbolic array ====>	1:	# of columns

```
« OBJ→ OVER SIZE
  → row col
  « 1 row
    FOR i i 1 - col * row + i - 1 + ROLL OBJ→ DROP
    NEXT row col
  »
»
```

→SM   Assemble Symbolic Array   (158)

106.5 bytes   #EBA7h

mn + 2:	... elements		mn + 2:	...
3:			3:	
2:	# of rows		2:	# of rows
1:	# of columns	====>	1:	# of columns

```
«  → row col
   «  1 row
      FOR i col →LIST col row i - * i + ROLLD
      NEXT row →LIST
   »
»
```

SMADD   Symbolic Matrix Addition   (158)

163 bytes   #46EEh

2:	symbolic matrix 1		2:	
1:	symbolic matrix 2	====>	1:	SM1 + SM2

```
«  -3 CF SWAP DUP DUP SIZE SWAP 1 GET SIZE
   → a2 a1 n m
   «  1 n
      FOR i 1 m
         FOR j a1 i GET j GET a2 i GET j GET + COLCT
         NEXT m →LIST
      NEXT n →LIST
   »
»
```

SMINV   Invert Symbolic Square Matrix   (159)

63 bytes   #60B2h

1:	square symbolic matrix	====>	1: inverse of matrix

```
«  DUP SCFACTR STRN SWAP SDET INV SMSMULT »
```

SMMULT   Symbolic Matrix Multiplication   (158)

216 bytes   #B372h

2:	symbolic matrix 1		2:	
1:	symbolic matrix 2	====>	1:	SM1*SM2

```
«  -3 CF DUP2 DUP SIZE SWAP 1 GET SIZE ROT
   DUP SIZE SWAP 1 GET SIZE
   → a1 a2 n2 m2 n1 m1
```

```
«  1 n1
   FOR i 1 m2
      FOR j 0 1 m1
         FOR k a1 i GET k GET a2 k GET j GET * +
         NEXT
      NEXT m2 →LIST
   NEXT n1 →LIST
»
```

SMSMULT   **Symbolic Scalar Multiplication**   (158)

167 bytes   **#EB2Ah**

2:	symbolic array or scalar		2:	
1:	scalar or symbolic array	====>	1:	s*SM

```
«  -3 CF
   IF DUP TYPE 5 ==
   THEN SWAP
   END
   → z
   «  DUP SIZE SWAP 1 GET SIZE
      → a n m
      «  1 n
         FOR i 1 m
            FOR j a i GET j GET z *
            NEXT m →LIST
         NEXT n →LIST
      »
   »
»
```

SMSOLV   **Symbolic System Solution**   (159)

75.5 bytes   **#795Ah**

3:	symbolic coefficients array		3:	
2:	symbolic constants array	====>	2:	
1:	list of variables		1:	list of solutions

```
«  → a b v
   «  v a SMINV b SMMULT →TAG
   »
»
```

SMSUB
Symbolic Matrix Subtraction (158)

163 bytes   #AE25h

2:	symbolic array 1		2:	
1:	symbolic array 2	====>	1:	

```
« -3 CF SWAP DUP DUP SIZE SWAP 1 GET SIZE
  → a2 a1 n m
  « 1 n
    FOR i 1 m
        FOR j a1 i GET j GET a2 i GET j GET - COLCT
        NEXT m →LIST
    NEXT n →LIST
  »
»
```

SNCOL
Insert Column in Symbolic Array (159)

77.5 bytes   #BC65h

3:	symbolic array		3:	
2:	symbolic column (list)		2:	
1:	column number	====>	1:	modified array

```
« → s v n
  « s STRN v n SNROW STRN
  »
»
```

SNROW
Insert Row in Symbolic Array (159)

76.5 bytes   #C674h

3:	symbolic array		3:	
2:	symbolic row (list)		2:	
1:	row number	====>	1:	modified array

```
« → s v n
  « s OBJ→ v OVER 3 + n - ROLLD 1 + →LIST
  »
»
```

1:	====>	1: { a b c A° B° C° area }

```
«
   «  → r s R
      «  IF R DUP SIN ASIN ≠ R SIN 1 == OR
         THEN
               IF r s >
               THEN r s R ←P2 →NUM
               ELSE 1 SF
               END
         ELSE
               IF r s ≥
               THEN r s R ←P2 →NUM
               ELSE
                  IF R SIN s * r − DUP ABS 1E-6 <
                  THEN DROP r s R ←P2 →NUM
                  ELSE
                        IF 0 >
                        THEN 1 SF 0
                        ELSE r s R ←P2 →NUM 3 SF
                        END
                  END
               END
         END
      »
   »
   «  → r s R « R SIN s * r / ASIN » »
   «  → r R S « S SIN r * R SIN / » »
   «  → r s T « r SQ s SQ + 2 r s * * T COS * − √ » »
   «  → r s t « r SQ s SQ − t SQ − 2 s * t * NEG / ACOS »
   »
   «  → R S « 180 R S + − » »
   «  → r s « a4 2 * r s * / ASIN » »
   «  → r S « a4 2 * r / S SIN / » »
   «  → R S « 180 R S + − SIN 2 * a4 * R SIN S SIN * / √
   » »
   «  → s t R « s t * R SIN * 2 / 'a4' STO » »
   { { 14 21 29 30 35 39 43 46 51 53 59 61 62 85 93 94 99
       107 110 111 115 117 123 125 126 } { 25 26 27 41 44
       45 50 52 54 57 58 60 89 90 105 108 109 114 116 118
       121 124 } { 28 42 49 92 106 113 } { 11 13 19 22 37
       38 } { 7 15 23 31 47 55 63 71 79 87 95 103 119 127 }
       { 67 69 70 75 77 78 83 86 91 101 102 } { 74 76 81 84
       97 98 } { 88 104 112 120 }
   }
   « 1 « "a" SWAP + OBJ→ » DOLIST SWAP 1 « "d" SWAP + OBJ→
   » DOLIST SWAP + DUP fields STO ←new SWAP STO
   »
   «  IF DEPTH DUP ROT →NUM DEPTH 1 − ROT ==
      THEN { NOVAL } 1 GET
      END SWAP DROP
   »
   «  1 3
```

```
        FOR j
          IF ←angles j GET 0 == ←sides j GET 0 ≠ AND
          THEN ←angles j 3 PUT ←sides j 3 PUT '←sides' STO
             '←angles' STO
          END
        NEXT
        IF ←angles 3 POS 0 ==
        THEN ←angles DUP 0 POS DUP 3 ROLLD 3 PUT ←sides ROT
           3 PUT '←sides' STO '←angles' STO
        END 1 3
        FOR j
          IF ←angles j GET 3 ≠
          THEN j
          END
        NEXT DUP ←angles SWAP 2 PUT ←sides ROT 2 PUT 3 PICK 1
        PUT 3 ROLLD SWAP 1 PUT 4 4 PUT ←prep EVAL d1 d2 a3
        ←p4 →NUM DUP 'd3' STO d1 a3 ←p2 →NUM DUP 'a1' STO a3
        ←p6 →NUM 'a2' STO d1 d2 a3 ←p10 EVAL
    »
    «   ←sides DUP 0 POS DUP 3 ROLLD 1 PUT ←angles ROT 1 PUT
        DUP 0 POS DUP 3 ROLLD 2 PUT 3 ROLLD 2 PUT DUP 14 POS
        DUP 3 ROLLD 3 PUT 3 ROLLD 3 PUT 4 4 PUT ←prep EVAL a1
        a2 ←p6 →NUM DUP 'a3' STO d1 a1 a2 ←p3 →NUM DUP 'd2'
        STO d1 ROT ←p4 →NUM DUP 'd3' STO d2 a1 ←p10 EVAL
    »
    «   1 3
        FOR j
          IF ←sides j GET 0 == ←angles j GET 0 ≠ AND
          THEN ←sides j 3 PUT ←angles j 3 PUT '←angles' STO
             '←sides' STO
          END
        NEXT
        IF ←sides 3 POS 0 ==
        THEN ←sides DUP 0 POS DUP 3 ROLLD 3 PUT ←angles ROT 3
          PUT '←angles' STO '←sides' STO
        END 1 3
        FOR j
          IF ←angles j GET 3 ≠
          THEN j
          END
        NEXT DUP ←sides SWAP 2 PUT ←angles ROT 2 PUT 3 PICK 1
        PUT 3 ROLLD SWAP 1 PUT SWAP 4 4 PUT ←prep EVAL d3 a1
        a2 ←p6 →NUM DUP 'a3' STO a1 ←p3 →NUM DUP 'd1' STO d3
        a2 ←p4 →NUM 'd2' STO d1 d2 a3 ←p10 EVAL
    »
    «   ←angles REVLIST TAIL REVLIST 0 POS ←angles SWAP DUP 3
        ROLLD 1 PUT ←sides ROT 1 PUT DUP 0 POS DUP 3 ROLLD 2
        PUT 3 ROLLD 2 PUT DUP 14 POS DUP 3 ROLLD 3 PUT 3
        ROLLD 3 PUT SWAP 4 4 PUT ←prep EVAL d1 d2 a1 ←p1
        EVAL
        IF DUP
        THEN DUP 'a2' STO a1 ←p6 →NUM DUP 'a3' STO d1 d2 ROT
          ←p4 →NUM DUP 'd3' STO d1 a2 ←p10 EVAL
        END
    »
    «   ←sides 1 { 1 2 3 } REPL ←angles 1 { 1 2 3 4 } REPL
        ←prep EVAL d1 d2 d3 ←p5 →NUM DUP 'a1' STO d1 d2 ROT
        ←p2 →NUM DUP 'a2' STO a1 ←p6 →NUM DUP 'a3' STO d1 d2
```

```
        ROT ←p10 EVAL
  »
  «  1 2
     FOR j ←sides DUP 0 POS DUP 3 ROLLD j PUT ←angles ROT
        j PUT '←angles' STO '←sides' STO
     NEXT ←sides DUP 14 POS DUP 3 ROLLD 3 PUT ←angles ROT
     3 PUT 4 4 PUT ←prep EVAL d1 d2 ←p7 →NUM DUP 'a3' STO
     d1 d2 ROT ←p4 →NUM DUP 'd3' STO d1 d2 ROT ←p5 →NUM
     DUP 'a1' STO a3 ←p6 →NUM 'a2' STO
  »
  «  ←sides DUP 0 POS DUP 3 ROLLD 1 PUT ←angles ROT 1 PUT
     DUP 0 POS DUP 3 ROLLD 2 PUT 3 ROLLD 2 PUT DUP 14 POS
     DUP 3 ROLLD 3 PUT 3 ROLLD 3 PUT 4 4 PUT ←prep EVAL d1
     a2 ←p8 →NUM DUP 'd3' STO d1 a2 ←p4 →NUM DUP 'd2' STO
     d1 d3 ROT ←p5 →NUM DUP 'a1' STO a2 ←p6 →NUM 'a3' STO
  »
  «  1 2
     FOR j ←angles DUP 0 POS DUP 3 ROLLD j PUT ←sides ROT
        j PUT '←sides' STO '←angles' STO
     NEXT ←sides DUP 14 POS DUP 3 ROLLD 3 PUT ←angles ROT
     3 PUT 4 4 PUT ←prep EVAL a1 a2 ←p6 →NUM DUP 'a3' STO
     a1 a2 ←p9 →NUM DUP 'd3' STO SWAP a2 ←p3 →NUM DUP 'd2'
     STO d3 a1 ←p4 →NUM 'd1' STO
  »
  «  "SOLVE TRIANGLE" { {"a:" "ENTER SIDE A" 0 9 } { "b:"
     "ENTER SIDE B" 0 9 } { "c:" "ENTER SIDE C" 0 9 } {
     "A":" "ENTER ANGLE A IN DEGREES" 0 9 } { "B":" "ENTER
     ANGLE B IN DEGREES" 0 9 } { "C":" "ENTER ANGLE C IN
     DEGREES" 0 9 } { "AREA:" "ENTER AREA OF TRIANGLE" 0 9
     } { } } { 2 2 } { } fields 1 ←nv DOLIST INFORM
  »
  RCLF
  →  ←p1 ←p2 ←p3 ←p4 ←p5 ←p6 ←p7 ←p8 ←p9 ←p10 ←cases ←prep
     ←nv ←sas ←aas ←asa ←ssa ←sss ←kss ←ksa←kaa ←infm flags
  «  DEG STD { 1 2 3 } CF { a b c A" B" C" K } DUP 1
     « PGALL » DOLIST 'fields' STO { NOVAL } DUP DUP + DUP
     DUP + + + fields STO
     WHILE ←infm EVAL
     REPEAT CLLCD "Solving triangle ..." 1 DISP 1
        «  IF DUP TYPE 9 ==
           THEN →NUM
           END
        » DOLIST DUP 1 « TYPE » DOLIST DUP 1 3 SUB SWAP 4
        7 SUB
        →  ←new ←sides ←angles
        «  0 1 7
           FOR j
              IF ←new j GET TYPE 0 ==
              THEN j 1 - 2 SWAP ^ +
              END
           NEXT 'case' STO ←cases 1
           « IF case POS
             THEN 1
             ELSE 0
             END
           » DOLIST 1 POS
           IF DUP
```

```
              THEN { d1 d2 d3 a1 a2 a3 a4 } PURGE { ←sas
              ←aas ←asa ←ssa ←sss ←kss ←ksa ←kaa } SWAP GET
              EVAL EVAL fields 1 ←nv DOLIST fields STO
              IF 1 FS?C
              THEN "No solution exists" MSGBOX ←new fields
                  STO
              ELSE 1 7
                  FOR j
                      IF ←new j GET { NOVAL } 1 GET ≠
                      THEN
                          IF fields j GET →NUM ←new j GET -
                            ABS 1E-6 >
                          THEN 2 SF
                          END
                      END
                  NEXT
                  IF 2 FS?C
                  THEN "No solution exists" MSGBOX ←new
                      fields STO
                  END
              END
              IF 3 FS?C
              THEN "One of two solutions" MSGBOX
              END
          ELSE "Not enough information" MSGBOX ←new fields
              STO
          END
        »
    END fields 1 ←nv DOLIST flags STOF {fields case a b c
    A" B" C" K d1 d2 d3 a1 a2 a3 a4 } PURGE
  »
»
```

SPIRO   Spirograph™ Simulation   (84)

411 bytes   #A325h

3:	number of teeth in fixed wheel		3:
2:	number of teeth in rolling wheel		2:
1:	-1 if inside roll; 1 if outside roll	====>	1:

```
« -3 CF DUP2 * 4 PICK +
  → a b s n
  « 'θ' PURGE n θ COS * b 1.5 * n b / θ * COS * s * - n
    θ SIN * b 1.5 * n b / θ * SIN * - i * + STEQ
    PARAMETRIC RAD 'θ' 0 8 FIX a b / →Q STD OBJ→ DROP2
    SWAP DROP '2*π' * →NUM 3 →LIST INDEP .05 RES
    IF s 1 ==
    THEN a ABS b ABS 3 * +
    ELSE a ABS b ABS 1.5 * +
    END DUP NEG SWAP DUP2 YRNG 2 * SWAP 2 * SWAP XRNG
    ERASE DRAW PICTURE
  »
»
```

sqrt

Square Root UDF (297)

37.5 bytes #9732h

1:	real number	====>	1:	square root

```
« → x « x √ » »
```

SRCI

RCI for Symbolic Matrix (158)

75 bytes #6D06h

3:	symbolic array		3:	
2:	multiplicative factor		2:	
1:	row	====>	1:	modified array

```
« -3 CF
  → s f n
  « s DUP n GET f * n SWAP PUT
  »
»
```

SRIJ

RCIJ for Symbolic Matrix (159)

98.5 bytes #5A32h

4:	symbolic array		4:	
3:	multiplicative factor		3:	
2:	row i		2:	
1:	row j	====>	1:	modified array

```
« -3 CF
  → s f i j
  « s i GET f * s j GET ADD s j ROT PUT
  »
»
```

SRSWP

Swap Rows in Symbolic Matrix (158)

87.5 bytes #7BFEh

3:	symbolic array		3:	
2:	row 1		2:	
1:	row 2	====>	1:	modified array

```
« → s i j
  « s DUP i GET SWAP j GET s i ROT PUT j ROT PUT
  »
»
```

STRN

Transpose Symbolic Matrix (158)

124 bytes   #8AD8h

1:	symbolic matrix	====>	1:	transposed symbolic matrix

```
«  DUP DUP SIZE SWAP 1 GET SIZE
   → a n m
   «  1 m
      FOR j 1 n
         FOR i a i GET j GET
         NEXT
      NEXT m n
   »  →SM
»
```

SXCOL

Extract Column from Symbolic Matrix (159)

73.5 bytes   #4706h

2:	symbolic array		2:	reduced array
1:	column number	====>	1:	extracted column (list)

```
«  → s n
   «  s STRN n SXROW SWAP STRN SWAP
   »
»
```

SXROW

Extract Row from Symbolic Matrix (159)

80 bytes   #A2F7h

2:	symbolic array		2:	reduced array
1:	row number	====>	1:	extracted row (list)

```
«  → s n
   «  s OBJ→ DUP 2 + n - ROLL OVER 1 + ROLLD 1 - →LIST
      SWAP
   »
»
```

SYND

Synthetic Division (input form) (117)

555.5 bytes   #FDDDh

2:	array		2:	
1:	program	====>	1:	modified array

```
«  «  IF DEPTH DUP ROT →NUM DEPTH 1 - ROT ==
      THEN { NOVAL } 1 GET
      END SWAP DROP
   »
```

```
«   "SYNTHETIC DIVISION"
    { { "POLYNOMIAL:" "ENTER POLYNOMIAL AS VECTOR" 3 4 }
    { "FACTOR:" "ENTER FACTOR TO BE TESTED" 0 1 }
    { "QUOTIENT:" "DISPLAYS COMPUTED QUOTIENT" }
    { "REMAINDER:" "DISPLAYS COMPUTED REMAINDER" } }
    1 { } fields 1 ←nv DOLIST INFORM
    IF
    THEN OBJ→ 3 DROPN
       → p f
       «  p f SDIV 4 →LIST 'fields' STO ←synr EVAL »
    END
» { NOVAL NOVAL NOVAL NOVAL } 'fields' STO
→ ←nv ←synr
«  ←synr EVAL » 'fields' PURGE
»
```

TNCON

Find Tangent and Normal at Point on Conic   (267)

328.5 bytes   #2B86h

2: {A B C D E F}		2: Normal: 'y=mx+b'
1: (x,y)	====>	1: Tangent: 'y=nx+d'

```
«  C→R
   →  con x1 y1
   «  RCLF -3 CF -22 SF 'x' PGALL con { 'x^2' 'x*y' 'y^2'
      x y 1 } * ΣLIST 'x' ∂ DUP x1 'x' STO EVAL
      →  fp fpx
      «  'y' fpx DUP INV NEG 2 →LIST DUP 'x' * SWAP x1 * -
         y1 ADD COLCT = OBJ→ DROP "Normal" →TAG SWAP
         "Tangent" →TAG ROT STOF
      »
   »
»
```

TRIGX

Trigonometry Explorer   (32)

2439.5 bytes   #A403h

1:		1:
	====>	

```
«  «  "TRIGONOMETRY EXPLORER" { { "∡(DMS)):" "ANGLE IN
   DD.MMSS" 0 } { "∡:" "ANGLE IN RADIANS" 0 9 } {
   "RADIUS:" "RADIUS OF CIRCLE" 0 9 } { "SIN:" "SINE OF
   ANGLE" 0 9 } { "ARC:" "LENGTH OF ARC INSCRIBED BY
   ANGLE" 0 9 } { "COS:" "COSINE OF ANGLE" 0 9 } {
   "AREA:" "AREA OF CIRCULAR SECTOR" 0 9 } { "TAN:"
   "TANGENT OF ANGLE" 0 9 } } { 2 2 } { } angd angr
   radi sine arc cosine area tang 8 →LIST INFORM
   »
   → ←infm
   «  RCLF { -2 -3 } CF { 45 'π/4' 1 '√2/2' 'π/4' '√2/2'
      'π/8' 1 } DUP 'old' STO { angd angr radi sine arc
      cosine area tang } DUP 'fields' STO STO
```

```
WHILE ←infm EVAL
REPEAT DUP fields STO
  → new
  «  { } 'inputs' STO 1 8
     FOR n
        IF new n GET DUP { NOVAL } 1 GET ≠ SWAP old
           n GET ≠ AND
        THEN fields n GET 'inputs' STO+
        END
     NEXT
     IF inputs { } SAME
     THEN new OBJ→ DROP
     ELSE 1 1 2
        FOR j
           IF inputs { angd angr } DUP j GET DUP 4
              ROLL SWAP POS
           THEN POS +
           ELSE DROP2
           END
        NEXT
        { «  1 1 4
             FOR j
                IF inputs { sine cosine 0 tang } DUP
                   j GET DUP 4 ROLL SWAP POS
                THEN POS +
                ELSE DROP2
                END
             NEXT RAD old 2 GET
             «  sine ASIN »
             «  cosine ACOS »
             «  'angr' DUP DUP SIN sine - SWAP COS
                cosine - = SWAP 'π/4' →NUM ROOT »
             «  tang ATAN »
             «  'angr' DUP DUP SIN sine - SWAP TAN
                tang - = SWAP 'π/4'
                   IF sine SIGN tang SIGN ≠
                   THEN 'π' +
                   END →NUM ROOT »
             «  'angr' DUP DUP COS cosine - SWAP TAN
                tang - = SWAP 'π/4'
                   IF cosine SIGN tang SIGN ≠
                   THEN 'π' +
                   END →NUM ROOT »
             DUP 8 →LIST SWAP GET EVAL DUP 'angr'
             STO →NUM R→D →HMS 'angd' STO
          »
          «  angd →NUM HMS→ D→R 'angr' STO »
          «  angr →NUM R→D →HMS 'angd' STO »
          «  IF -17 FS?
             THEN angr →NUM R→D →HMS 'angd' STO
             ELSE angd →NUM HMS→ D→R 'angr' STO
             END
          »
        } SWAP GET EVAL 1 1 4
        FOR j
           IF inputs { radi arc 0 area } DUP j
              GET DUP 4 ROLL SWAP POS
           THEN POS +
```

```
                    ELSE DROP2
                    END
                  NEXT
                  { radi « arc angr / » radi « area 2 * angr
                  / ABS √ » radi « arc angr / » radi
                  } old 3 GET SWAP + SWAP GET EVAL RAD angd
                  angr ROT OVER SIN 3 PICK 3 PICK * 4 PICK COS
                  5 PICK 5 PICK SQ * 2 / ABS 6 PICK TAN
              END 8 →LIST 1
              « →NUM 10 FIX →Qπ
                IF DUP TYPE
                THEN →RPN
                  IF DUP { / } 1 GET POS 0 ≠
                  THEN
                      IF DUP 2 GET 100 ≤
                      THEN EVAL →Qπ
                      ELSE EVAL →NUM
                      END
                  ELSE EVAL
                  END
                END STD
              » DOLIST DUP fields STO 'old' STO
            »
          END STOF { angd angr radi sine arc cosine area tang
          fields inputs old } PURGE
        »
      »
    »
```

TRNCON
Translate Conic (272)

249.5 bytes   # 8FBh

2: {A B C D E F}		2:
1: [x y]	====>	1: {A' B' C' D' E' F'}

```
« OBJ→ DROP ROT OBJ→ DROP
  → h k a b c d e f
  « a b c b d + 2 h a * * - e b h * - 2 k c * * - a h
    SQ * b h k * * + c k SQ * + d h * - e k * - f + 6
    →LIST
  »
»
```

TYIEW
View Transformed Array of Points (214)

370 bytes   #184Ah

1: object array of points	====>	1: object array of points

```
« { PICT PPAR } PURGE ERASE DUP SIZE OBJ→ DROP (0,0)
  → a m n s
  « a DUP { m n } GET / 'a' STO 1 m
    FOR k a k ROW- 1 2 SUB V→ R→C DUP s + 's' STO SWAP
      DROP
```

```
          NEXT m →LIST 'p' STO s m / DUP (6,3) - SWAP (6,3) +
          PDIM 1 m
          FOR j p j GET p j m MOD 1 + GET LINE
          NEXT 'p' PURGE { } 1 ATICK DRAX PVIEW a
      »
  »
```

UDFUI

Apply User Interface to a UDF (27)

359 bytes   #9C2Ah

1:	name of UDF	====>	1:	

```
«  { }
   → name infdat
   «  name RCL →STR 4 OVER SIZE SUB "'«" + 1 OVER "'" POS 3
      PICK "«" POS MIN 2 - SUB "{" SWAP + OBJ→ "" name +
      SWAP name + 1 « ":" + » DOLIST DUP SIZE 4 / CEIL OVER
      1 « DROP { NOVAL } HEAD » DOLIST DUP 5 ROLLD 4 →LIST
      'infdat' STO
      WHILE infdat OBJ→ DROP 5 ROLL INFORM
      REPEAT DUP OBJ→ DROP2 name EVAL DUP 3 PICK SWAP name
        →TAG 4 ROLLD SIZE SWAP PUT
      END
   »
»
```

VDIR

Find Vector Direction Angles (177)

101.5 bytes   #14F8h

1:	vector	====>	1:	list of direction angles

```
«  DUP SIZE 1 GET SWAP DUP ABS
   → n v m
   «  1 n
      FOR i v i GET m / ACOS
      NEXT n →LIST
   »
»
```

I. Indexes

Examples Index

1. Exploring Functions

2. Trigonometry

3. Polar and Parametric Equations

4. Polynomials

5. Systems of Linear Equations

6. Analytic Geometry

7. Conic Sections

Programs Index

Subject Index

Converting decimals to fractions   40
Converting one-variable function to polynomial   133
Converting one linear equation form to another
     193-195
Coordinate system   220
Coordinates
     display of   58-59
     homogeneous   213
     of graphics cursor   16
Coplanar   201, 209, 250
Cosecant   29
Cosine   29
Cotangent   29
Cramer's Rule   144, 150–152, 159
Cross product   174-175, 187- 188, 190, 196-198,
     203, 205, 208, 211
Curtate curves   84-85, 88, 91-93
Curves
     Cassinian   76–78
     cissoids and conchoids   79–83
     cycloidal   84–93
     spirals   94–97
Cycloid   84
Cycloidal curves   84–93

Defining procedure in user-defined functions   24
Degrees   30–31, 51
Deltoid   90
Depressed polynomial   121
Derivatives   104-107
Descartes' Rule of Signs   119-121, 124-125
Determinant 142, 144, 150–155, 158
Diagonal elements of a matrix   143, 214
Dimensions of a matrix   139–141
Dimetric projection   212, 228, 231-232
Direction angles of a vector   176-177
Direction vectors of lines   185, 192, 196-197, 202,
     205, 209-210, 225
Directory   27, 275
Directrix   243, 257, 259, 263
Discontinuities   18
Discriminant of a quadratic   13
Display range   26, 31, 39-40, 61, 247
Distance
     between center and focus   261
     between focus and vertex   259
     between parallel lines   179, 201
     between two points   178, 180
     from a point to a line   178-179, 189, 198-199
     from a point to a plane   179, 207
Division
     of matrices   143
     of polynomials   112–113
     synthetic   115–124
Domain of a function   11
Dot product   140, 174-175, 187-188, 202, 208-211

Eccentricity   243, 255, 257, 259, 261, 263- 264
Ellipse   243, 246, 255–258, 269, 271
     angle of orientation   255-256, 258
     center of   255-256, 258
     directrixes   257-258
     eccentricity   255-258
     foci   255, 257-258
     semimajor   255-258
     semiminor   255-256, 258
     vertices   257-258
Epicycloid   69, 84, 86-87
Epitrochoid   88-89
EQ variable   124
Equation of a circle   250, 252-253, 268-269
Equation of a hyperbola 261-264
Equation of a line
     array form   189, 193-195, 199-201
     determined by three points   178
     determined by two points   178, 182
     general form   178, 189, 193-195, 199-201
     intersection of two planes   179
     parametric form   178, 183, 185, 192-201, 203,
          205-206, 210-211, 225
     perpendicular to a line 197
     perpendicular to two other lines   179, 188
     position-direction (PD) form   193-195, 197-199,
          201, 203, 209-211, 225
     slope-intercept form   178, 182, 192-195, 265
Equation of a parabola   259-260
Equation of a plane
     determined by a point and a line   179, 196
     determined by a point and plane 179, 208
     determined by its normal and a point   179
     determined by its traces   179, 204
     determined by three points   178, 187-188
     determined by two noncollinear lines   179, 203
     general form   187, 205, 207
     vector form   206, 211
Equation of an ellipse   255, 258
Equation of general conic   243, 248, 250, 259
Equation of the normal to a plane   179
Equations
     consistent vs. inconsistent   135
     degenerate   135
     independent   135, 137
EquationWriter   36
Equiangular spiral   95
Euclidean norm   160
EXPAN command   14
Expansion of a polynomial   114
Exponential functions   20–21
Extracting rows or columns of a matrix   159
Extrema   104-105
Eyepoint   228, 234, 236, 238-241

length 190
 midpoint of 178, 181, 184-185
 perpendicular bisector of 178, 184-185, 227, 252
Linear equations 134
 as a matrix equation 153
 solving 145–156, 205
 symbolic solutions 157–159
Linear inequalities 161–171
Linear programming 162–171
Lines 135
 collinear 179
 concurrent 135, 179, 200
 direction vectors of 185, 192, 197, 205
 in space 183, 185, 192-193, 198-199, 223
 intersecting 135, 200, 202-203
 parallel 179, 200-203, 212
 perpendicular 179
 skew 179, 200-201
 slope-intercept form 9, 192-193
Lissajous 98
Lists 13
Lituus spiral 96
Local names in user-defined functions 24
Logarithmic functions 20–21
Logarithmic spiral 95
Logarithms 21
Long division 115, 116
Lower bound of the roots of a polynomial 119-120, 122-124
Lower Bound Theorem 119

Matrices
 addition of 139, 158, 174
 and vectors 173
 arithmetic with 139–143, 158
 assembling 158
 augmented 145, 148-150, 152, 157, 159
 characterizing 135
 cofactor 153-155, 159
 condition number of 136-138
 determinant of 142, 150–152, 154-155, 158
 dimensions of 139
 disassembling 158
 editing 138
 extracting rows or columns 159
 identity 143, 145, 149, 214, 234
 ill-conditioned 137
 inserting rows or columns 159
 inverse of 142-144, 149, 153–156, 159
 multiplication 140–141, 153, 158, 175, 216, 222-223, 226, 231
 rank of 136, 138
 reduced row echelon 145, 147-149
 representing linear programs 162
 row echelon 145, 147
 row operations 159

row operations on 144-145, 148, 158
 scalar multiplication 139, 158
 subtraction 140, 158, 174
 symbolic 143, 157–159
 transformation 213–241
 transposing 141, 154-155, 158
Matrix equations 153
Maxima 102, 104
MaxMin problem 170-171
Median of a triangle 190-191
Midpoint of a line segment 178, 181, 184-185, 191, 261
Minima 102, 104
MinMax problem 171
Multiplication
 matrices 140–141, 153, 158, 175, 216, 222-223, 226, 231
 scalars and matrices 139, 158, 174

n1 variable 50
Names 24
Nephroid 87
Normal to a conic 267–273
Normal to a plane 179, 187- 188, 205-210
Normalization of a transformation 234-236
Numerical precision 39-40, 167

Objective function in a linear program 162-166, 168, 170-171
Optimization 162–171
Orthographic projections 228-230
Ovals of Cassini 76

Parabola 243, 259–260, 269, 272
 angle of orientation 259-260
 directrix 259-260
 eccentricity 259
 focus 259
 p parameter 260
 vertex 259-260
Parabolic axis 260
Parabolic spiral 96
Parallel lines 179, 197, 200-201, 212, 228
Parallel planes 187, 205-206, 208, 230
Parameters 57, 68
 angle 57
 of a vector 176
 time 57, 67, 70
Parametric functions 57-58
 plotting 66–71
Parametric plot type 72–75
 and complex numbers 66
Pascal's Snails 82
Path 27, 275
Pattern matching 36, 40
Perimeter of a triangle 190

If you liked this book, there are others that you will certainly enjoy also:

An Easy Course in Programming the HP 48G/GX

Here is an *Easy Course* in true Grapevine style: Examples, illustrations, and clear, simple explanations give you a real feel for the machine and how its many features work together. First you get lessons on using the Stack, the keyboard, and on how to build, combine and store the many kinds of data objects. Then you learn about programming—looping, branching, testing, etc.—and you learn how to customize your directories and menus for convenient "automated" use. And the final chapter shows example programs—all documented with comments and tips.

Calculus on the HP 48G/GX

Get ready now for your college math! Plot and solve problems with this terrific collection of lessons, examples and program tricks from an experienced classroom math teacher:

Limits, series, sums, vectors and gradients, differentiation (formal, stepwise, implicit, partial), integration (definite, indefinite, improper, by parts, with vectors), rates, curve shapes, function averages, constraints, growth & decay, force, velocity, acceleration, arcs, surfaces of revolution, solids, and more.

Graphics on the HP 48G/GX

Here's a must-have if you want to use the full potential of that big display. Written by HP engineer Ray Depew, this book shows you how to build graphics objects ("grobs") and use them to customize displays with diagrams, pictures, 3D plots—even animation.

First the book offers a great in-depth review of the built-in graphics tools. Then you learn how to build your own grobs and use them in programs—with very impressive results!

The HP 48G/GX Pocket Guide

You get some 90 pages of quick-reference tables, diagrams, and handy examples—all in a convenient little booklet that *fits in the case with your HP 48G/GX!* There is a complete command reference, along with system flags, menus, application summaries, troubleshooting, and common Q's & A's. Nothing is more succinct and helpful than this little memory-jogger!

For more details on these books or any of our titles, check with your local bookseller or calculator/computer dealer. Or, for a full Grapevine catalog, write, call or fax:

Grapevine Publications, Inc.
626 N.W. 4th Street   P.O. Box 2449
Corvallis, Oregon 97339-2449 U.S.A.
Phone: 1-800-338-4331 *or* 503-754-0583
Fax: 503-754-6508

ISBN		Price*
	Books for personal computers	
0-931011-28-0	**Lotus** Be Brief	$ 9.95
0-931011-29-9	A Little **DOS** Will Do You	9.95
0-931011-32-9	Concise and **WordPerfect**	9.95
0-931011-37-X	An Easy Course in Using **WordPerfect**	19.95
0-931011-38-8	An Easy Course in Using **LOTUS 1-2-3**	19.95
0-931011-40-X	An Easy Course in Using **DOS**	19.95
	Books for Hewlett-Packard Scientific Calculators	
0-931011-18-3	An Easy Course in Using the **HP-28S**	9.95
0-931011-25-6	**HP-28S** Software Power Tools: **Electrical Circuits**	9.95
0-931011-26-4	An Easy Course in Using the **HP-42S**	19.95
0-931011-27-2	**HP-28S** Software Power Tools: **Utilities**	9.95
0-931011-31-0	An Easy Course in Using the **HP 48S/SX**	19.95
0-931011-33-7	**HP 48S/SX** Graphics	19.95
0-931011-XX-0	**HP 48S/SX** Machine Language	19.95
0-931011-41-8	An Easy Course in Programming the **HP 48G/GX**	19.95
0-931011-42-6	Graphics on the **HP 48G/GX**	19.95
0-931011-43-4	Algebra and Pre-Calculus on the **HP 48G/GX**	19.95
0-931011-44-2	The **HP 48G/GX** Pocket Guide *(Available 11/94)*	9.95
0-931011-45-0	Calculus on the **HP 48G/GX** *(Available 1/95)*	19.95
	Books for Hewlett-Packard Business calculators	
0-931011-08-6	An Easy Course in Using the **HP-12C**	19.95
0-931011-12-4	The **HP-12C** Pocket Guide: **Just In Case**	6.95
0-931011-19-1	An Easy Course in Using the **HP 19Bɪɪ**	19.95
0-931011-20-5	An Easy Course In Using the **HP 17Bɪɪ**	19.95
0-931011-22-1	The **HP-19B Pocket Guide:** Just In Case	6.95
0-931011-23-X	The **HP-17B Pocket Guide:** Just In Case	6.95
0-931011-XX-0	**Business Solutions** on Your HP Financial Calculator	9.95
	Books for Hewlett-Packard computers	
0-931011-34-5	**Lotus in Minutes** on the **HP 95LX**	9.95
0-931011-35-3	**The Answers You Need** for the **HP 95LX**	9.95
	Other books	
0-931011-14-0	**Problem-Solving Situations:** A Teacher's Resource Book	9.95
0-931011-39-6	**House-Training Your VCR:** A Help Manual for Humans	9.95

Contact: **Grapevine Publications, Inc.**
626 N.W. 4th Street P.O. Box 2449 Corvallis, Oregon 97339-2449 U.S.A.
800-338-4331 (503-754-0583) *Fax:* 503-754-6508

*Prices shown are as of 8/6/94 and are subject to change without notice. Check with your local bookseller or electronics/computer dealer—or contact Grapevine Publications, Inc.

Reader Comments

We here at Grapevine like to hear feedback about our books. It helps us produce books tailored to your needs. If you have any specific comments or advice for our authors after reading this book, we'd appreciate hearing from you!

Which of our books do you have?

Comments, Advice and Suggestions:

May we use your comments as testimonials?

Your Name: Profession:
City, State:

How long have you had your calculator?

Please send Grapevine catalogs to these persons:

Name _____

Address _____

City _____ State _____ Zip _____

Name _____

Address _____

City _____ State _____ Zip_____